T0321062

Off-label prescribing – Justifying unapproved medicine

Off-label prescribing – Justifying unapproved medicine

David Cavalla

WILEY Blackwell

This edition first published 2015 © 2015 by John Wiley & Sons, Ltd.

Registered Office
John Wiley & Sons, Ltd., The Atrium, Southern Gate, Chichester, West Sussex,
PO19 8SQ, UK

Editorial Offices
9600 Garsington Road, Oxford, OX4 2DQ, UK
The Atrium, Southern Gate, Chichester, West Sussex, PO19 8SQ, UK
111 River Street, Hoboken, NJ 07030-5774, USA

For details of our global editorial offices, for customer services and for information about how to apply
for permission to reuse the copyright material in this book please see our website at
www.wiley.com/wiley-blackwell.

The right of the author to be identified as the author of this work has been asserted in accordance with the
UK Copyright, Designs and Patents Act 1988.

Library of Congress Cataloging-in-Publication Data

Cavalla, David, author.
  Off-label prescribing : justifying unapproved medicine / David Cavalla.
    p. ; cm.
  Includes bibliographical references and index.
  ISBN 978-1-118-91207-2 (cloth)
  I. Title.
  [DNLM: 1. Off-Label Use.   2. Drug Approval.   3. Pharmaceutical Preparations.   QV 748]
  RM300
  615.1–dc23
                          2014028488

A catalogue record for this book is available from the British Library.

Wiley also publishes its books in a variety of electronic formats. Some content that appears in print may not
be available in electronic books.

Cover image: © Dan Jubb
Cover design by Dan Jubb

Set in 9/11pt Meridien by SPi Publisher Services, Pondicherry, India

1   2015

This book is for patients, who for too long have been misled about the fact that many of the prescriptions that are written for them are for unapproved for their circumstance. We are all patients, or potential patients, so really this book is for everyone.

Contents

Foreword

NORD (The National Organisation for Rare Disorders) estimates that 25 000 000 Americans, 8% of the population, have a life-altering disease for which there is no currently effective therapy. Globally, 8% of the population would yield over 500 000 000 people similarly affected. Yet, under the current system, with all the knowledge, technology and money we have to invest in this problem, most people with diseases for which there are no, or only poor, treatment options have little hope of receiving an effective treatment in their lifetimes. Healthcare costs round the world keep rising, and a significant portion is spent on palliative care for diseases with no truly effective treatment. Those costs, plus lost productivity costs and the emotional trauma for patients and their families, directly or indirectly impact all of us.

The for-profit medical research industry is our current 'solution', but it can only work for some patients, and many serious conditions are left unaddressed. The major pharmaceutical companies invest more than US$ 70 billion per year in R&D to bring to market about 30 new drugs or drug improvements annually. Development takes 10–14 years and costs US$ 1.5 billion or more per new drug. Industry generally makes a lower investment in rare diseases, acute diseases, prevention and diseases of the poor where it cannot make a suitable profit. This is essentially a market failure, which restricts many patients from receiving solutions to their medical problems, and makes it unlikely that the for-profit system can conquer most of the 7000 diseases waiting to be addressed.

Other factors compound the problem. Academic research for diseases of the poor and rare diseases receive limited funding. Researchers often cannot or would not collaborate due to intellectual property and authorship concerns, so the limited funds that are available are not leveraged by collaboration. And philanthropic and venture funders are often stymied in their efforts to find the best treatment ideas and creating the research partnerships required to create treatments for these underserved patients and diseases.

Physicians, patients, payers, government and industry are all searching for solutions to this gaping treatment hole. One stopgap measure employed with regularity around the globe is to use drugs approved for one disease to treat another disease for which formal approval has not been obtained: this is called 'off-label' medicine. While on the surface repurposing of our existing therapeutic armoury has great appeal, when one examines this in more detail, significant peril is exposed. In practice, the freedom to prescribe off-label has often been abused by prescribers and industry: products have been used with inadequate evidence for trivial conditions, and commercial interests have trumped patient welfare. In order to sort this out, we need to differentiate the acceptable off-label uses from the unacceptable. But how?

David Cavalla examines, in great detail and with clear support, the issues of off-label drug prescribing. His evaluation is both broad and deep. He notes the value and the pitfalls of the practice, and offers cogent and feasible solutions to create greater value for patients. Most importantly, while sharing his expertise, he gives the reader the chance to draw his or her own conclusions. This is a very important book, because catastrophic diseases do or will impact many of us. At some point in each of our lives, we are likely to be faced with the need to find a medical solution to an unresolved disease, either for ourselves or for someone we care about. And that solution might involve what David Cavalla calls 'An Unapproved Medicine'. Armed with the knowledge in this book, you might make a different set of decisions or make the same decision better informed.

Dr. Bruce E. Bloom
President and Chief Science Officer,
Cures Within Reach

Acknowledgement

In writing this book I would like to thank all of the people who have given me help and advice, including specifically my wife, my daughter Anna and father John, who read and reviewed the early draft of the manuscript. I am also most thankful for the expert advice from Drs David Erskine, Janice Steele and Michael Clementson from the delivery end of medicine, as pharmacists and practitioners.

Author's note on the cover design

Off-label medicine is the technical term for medicines which have not been approved for the therapeutic purpose for which they are prescribed. It is a term with which most patients are unfamiliar, yet it can be likened to something with which is much more recognisable: off-piste skiing. The likeness, depicted on the cover of this book, extends on the one hand to the fact that neither practice is strictly illegal, and on the other to the fact that both practices are less safe and well-described than the authorised alternatives. Off-label uses of medicines are not regulated, so we have much less information about the safety and efficacy of the treatments. But sometimes, like off-piste skiing, there is no other way to travel.

To justify the use of an off-label treatment, there is one and only one person to bear in mind: the patient. But disposing of the other interests in the delivery of medicine, for example the pharmaceutical company that makes the product, the doctor who prescribes it and the government or insurance company who pays for it, is not an easy task.

Introduction

When becoming authorised to practise, the Hippocratic Oath requires a doctor to swear to '...use treatments for the benefit of the ill...but from what is to their harm and injustice...[to]...keep them'[1].

It is common knowledge among the medical profession and the public at large that the Oath requires the doctor to essentially 'Do no harm'; however, the requirement to keep their patients from injustice is much less appreciated. This book will identify issues that relate to the justice of the relationship between doctor and patient, as well as a wider consideration of other aspects of the complex ways in which medicines travel from scientist's bench to the bedside.

Most importantly, it will deal with the way in which around one in five of prescriptions today are written outside regulatory purview – in other words, the treatments have not been approved by the regulatory agencies. This is what I mean by unapproved medicine. Given that there were over four billion prescriptions written in the United States in 2011 [2], it is a very significant issue.* In certain areas, the proportion of unapproved prescriptions is much higher even than this, reaching three quarters or even 90% of prescriptions in some types of patients or with certain conditions.

At this point, you, the Reader, will surely say: No, he is wrong. This cannot be. We have a highly regulated and legally constituted system by which the safety and efficacy of medicines is ensured before they are taken by patients. I do know that medicines sometimes have side effects, sometimes do not work and sometimes even the regulators get it wrong, but I simply cannot believe that prescriptions cannot be written without regulatory say-so. And on this scale, it beggars belief!

You will also say: How can I not have heard of this before? If it is true, why have the press not highlighted it more? What is the point of medicines regulation when nearly a quarter of prescriptions are not regulatorily approved? Being perhaps well read in this area, you may also say: I have heard there are issues of data being hidden from the public by pharmaceutical companies, but I thought companies were obliged to have all their products approved by regulators before they are dispensed. And, you go on: if this is so, why are regulators not more stringent with the rules, to prevent it happening?

If this is your response, please do read on, because what I say is true. And, for the most part, it is all perfectly legal. It is my intention that by the end of the book, readers will be able to judge whether, from the patient's perspective, our current

---

*Using data from the OECD (OECD Health Policy Studies Pharmaceutical Pricing Policies in a Global Market; OECD Publishing, 2008. DOI: 10.1787/9789264044159-en), the relative volume of pharmaceutical utilisation can be obtained across the OECD countries. This is then normalised according to the population and the known number of prescriptions in the United States in 2011, which is 4.02 billion according to Ref. [2]. The total number of prescriptions across the United States, Japan, France, Germany, Spain, the United Kingdom, Italy, Korea, Canada, Australia, Mexico, Poland, the Netherlands, Sweden, Portugal, Austria, Hungary, Czech Republic, Switzerland, Norway, Finland, Denmark, Ireland, Slovakia, New Zealand and Iceland is then estimated to be 10.85 billion. Twenty-one per cent of this is over two billion prescriptions per year, a number I shall refer to later in the book.

practice of medicine and its prescription meets the standard of justice espoused in the Hippocratic Oath.

This book explores the nooks and crannies of our medicated lives, where drug regulation runs up against medical practice, and concerns the use of a drug that has been approved for one use (in medical parlance, 'indication') being used for a different indication; alternatively, being used on a different set of patients from the ones it is approved for, or at a different dose. It is now time to shed some light on this somewhat dark area. As you will see, not only does this mean that the evidence base for the drug's benefit is suspect, but there are safety issues too. Usually the patient is unaware of what is going on, having not been informed by their doctor of this aspect of his or her prescribing choice. I will tell you what the various medical professions have to say about this, how they respond to regulatory bodies and how pharmaceutical companies benefit by moving into this poorly regulated area. The issues are complex and resist simplistic headline-grabbing sound bites; but I hope you will persist to the conclusion of this book, since, in addition to pointing out the problems, by the end I will also leave you with some proposals to improve the way medicines are prescribed and evidence gathered to support the ways they are used.

CHAPTER 1

What is off-label medication, and how prevalent is it?

The practice of medicine has been regulated since Hippocrates, who first told doctors (physicians, clinicians, general practitioners [GPs] and so on) how they should behave with regard to their patients. His Oath, written nearly 2500 years ago, is the most famous text in Western medicine. Though most people do not know exactly what it says, they believe it to say something along the lines of 'Doctor, do no harm'. That is only partly true, as I shall now explain.

But before I do, there are actually many versions in the public mind of what Hippocrates said, including the view recounted by one UK doctor of an elderly patient who believed the Oath instructed doctors never to tell patients the truth. This book will describe circumstances in which this patient is often correct, namely, that GPs do not tell the truth to their patients, but of course incorrect in that Hippocratic Oath does not say that.

The Oath starts: 'I swear by Apollo the physician and by Asclepius and Hygieia and Panacea... to bring the following oath to fulfilment'. According to Greek mythology, Apollo is the god of healing, Asclepius is his son and Hygieia and Panacea are his granddaughters. As with Zeus his father, Apollo had many love affairs with goddesses and mortals. One of his amours was Coronis, who was the daughter of the king of the Lapiths. Dwelling on a higher plane, Apollo was not able to be beside Coronis on earth, so he sent a white crow to look after her. Unfortunately, while she was pregnant by Apollo, Coronis fell in love with another man, and the crow informed Apollo of the affair. Appalled at her infidelity, in his anger, Apollo turned the crow black.

Artemis, Apollo's twin sister, shot an arrow to kill Coronis. While Coronis' body was burning on the funeral pyre, Apollo removed the unborn child, who was called Asclepius and became the god of medicine. When he grew up, Asclepius had two daughters, Hygieia, the goddess of health, and Panacea, the goddess of cures: medicine ran in the family. The words 'hygiene' and 'panacea' clearly have their etymological origins in these mythological figures.

According to legend, Hippocrates was a descendant of Asclepius; this gives more weight to Hippocrates' proclamations, particularly when he pronounces on medical matters. Part of the Oath instructs the doctor to treat his teachers as his parents and to pass on the art of medicine to the next generation of healers. This is clearly relevant to Hippocrates' ancestry, going all the way back to Apollo. But it is the next part of the Oath that is most relevant to this book and indeed to the practice of medicine.

It continues: 'And I will use treatments for the benefit of the ill in accordance with my ability and my judgment, but from what is to their harm and injustice I will keep them'.

Off-label Prescribing – Justifying Unapproved Medicine, First Edition. David Cavalla.
© 2015 John Wiley & Sons, Ltd. Published 2015 by John Wiley & Sons, Ltd.

It is the two words, 'harm' and 'injustice' which I ask you to bear in mind as we go forwards.

What is 'off-label' medicine?

Today, medicines are regulated for their efficacy and safety, and once licensed for sale, they can be marketed for certain uses as justified by the data. Regulatory bodies in developed countries are constituted by legal statute and operate as parts of government, ostensibly in the interests of the people as patients. But once approved, medicines can be used for any purpose the prescriber sees fit and appropriate for the patient. In other words, regulatory authorities are the gatekeepers to prevent the medical use of unapproved products, but then leave the gate entirely wide open regarding unapproved indications or uses of approved products. To be succinct, medicinal products require regulatory approval, but the practice of medicine does not. There remain restrictions on the marketing of these products, but these are considerations for the producer, not the prescriber. Later on, I will explain the nuance that distinguishes between the marketing and the use of medicines and how, in my opinion, pharmaceutical companies game the system.

The ways in which medicines are prescribed, and administered, outside the terms of the marketing authorisation are called 'off-label' uses. They have not been justified by the regulatory authorities, which determine the label for the product, hence the title of this book. As was said, a 'general off-label use of drugs is the death of the idea of regulation' [3]. The importance of the regulatory justification is not merely because these public authorities spend a lot of time, money and manpower examining the evidence behind the safety and efficacy of the medicines we take: it is because these authorities are put in place to implement certain standards to which the patient expects his or her therapy to accord. The regulatory approval is also the patient's approval, the basis for their consent to being treated with the prescribed medication. Drug regulation is a complex decision about the balance of safety and efficacy,[1] benefit and risk – a world of shades of grey, not black and white. In the real world, the prescribing doctor has a lot of flexibility as to what s/he can prescribe; that flexibility can be put to good use, but patients are rarely aware that their off-label medicine has not been approved for their affliction, with consequences to the quality of their care.

So, off-label prescriptions are not illegal, and from the doctor's perspective, they may not even be seen as unethical; in fact, according to the Hippocratic Oath, they may fulfil a doctor's moral imperative, for instance, in situations of rare diseases where there is no approved product. However, the evidence behind off-label medicine rarely fulfils the patient's expectations that a formal regulatory assessment of safety and efficacy has been performed, and this is the first sense in which I mean off-label medicine seems to be unjustified. Later, in Chapter 6, I shall deal with other consequences, such as who pays for the medicine, and what happens in cases where things go wrong. But before doing so, let us consider the scale of the issue.

There are lots of examples of secondary uses for existing drugs. The story of how a proposed treatment for angina and heart failure ended up as the world's first treatment for erectile dysfunction is well known. The company behind the drug (Pfizer), now known as Viagra™, recorded that when the product, then known as

---

[1]Clinicians and policymakers often differentiate between efficacy and effectiveness, where the latter relates to how well a treatment works in the practice of medicine, as opposed to the former, which measures how well treatment works in clinical trials or laboratory studies.

UK-92480-10, or sildenafil, was first tried on male volunteers in a Welsh clinic, they reported physical excitation on seeing the nurses in the ward, requiring them to roll on their stomachs. In this case, the intended development for cardiovascular diseases was curtailed, and the product entered into medical practice for the treatment of erectile dysfunction instead (and in 2012, generated over $2 billion in revenue for Pfizer). Because the decision to develop for erectile dysfunction occurred before Viagra was approved for any use, this is not an example of off-label medication. However, even though this story is somewhat anecdotal, it does show that drugs often do more than one thing. In fact, there is a sequel to the first approval indication for sildenafil, in which it was subsequently developed for a second indication (or third, depending on how you look at it), as we shall see in Chapter 2.

I have strong interest in this area, having investigated this area of secondary uses for existing drugs, now called drug repurposing, for over 15 years. I have collated over 2300 proposed new uses for existing drugs, either marketed products or investigational compounds. This is freely accessible on the internet at http://www.drugrepurposing.info. But the level of support for such new uses can vary enormously. In some cases, we have human data, such as clinical trials to support the effect. In many others, there is only information from experiments *in vitro* (literally 'in glass', this refers to test tube experiments) or *in vivo* (in animals). Some information even derives from a computer assessment of the shape similarity of drugs, but predictions like this based on *in silico* analysis are merely hypotheses, starting points for research programmes lasting years or even decades to deliver validation in regulatory studies that would be needed for market approval. As we shall discover in Chapter 4, most of the normal scientific hypotheses upon which drug discovery programmes are based turn out to be wrong.

We now realise that there are very few, if any, drugs with only one activity and/ or only one conceivable therapeutic use. But even though there is vast promise from making better use of the drugs we currently have on hand, most of the early-stage predictions fail to be realised in practice. Sometimes this is for commercial reasons, but it is also for experimental reasons of safety or efficacy. As this area becomes more widely used as a means to discover new therapeutics, it is all the more likely that the current medicines that we all use will become increasingly investigated for new uses. New discoveries of this kind can be enormously helpful to the armoury of therapeutics available to the patient. However, it is unsafe to suppose that a theory deriving from an animal experiment, or anecdotal case report from one patient, really translates into a safe, efficacious treatment of general merit: it needs to be proven. Prescribers have enormous freedom to uncover whether the early science suggestive of a human benefit really works in a patient. As this book will show, the current legal framework, regulatory controls and ethical norms in medicine do not provide the best environment for delivering such new therapeutics to patients, and the consequences of its misapplication can be gravely injurious.

There are two main types of off-label medicine: use of drugs for unapproved diseases or conditions (which, in the medical profession, are called 'indications'), and use of drugs for unapproved patient groups. Off-label use can also include prescribing different dosages, lengthening or shortening the interval between treatments or using different routes of administration from those indicated on the drug label.[2]

---

[2]A word about semantics. There is a difference between the terms 'unapproved drug' or 'unlicensed drug' and 'unapproved medicine'. The word 'drug' implies the active ingredient in the therapy, whereas 'medicine' connotes the entire formulation (including dose, frequency, etc.), its use and the type of patient.

There are three main areas of therapy where off-label medicines are most widely used. The first is the use of products licensed for adults, on the basis of clinical trials in adults, for children. The second is of psychiatric medicines, and the third is in oncology treatment. We started with a broad statement that off-label use constitutes '20% of all prescriptions', but the prevalence varies enormously, and among these broad classes lie salient examples where off-label use reaches staggering proportions. Getting consistent statistics can be difficult: a review of international studies in ambulatory care reports rates of 13.2% and 29%, in paediatric wards between 18% and 60% and in neonatal units between 14% and 63% [4]. Another international literature review reports that rates for off-label medicine use vary between 11% and 80% [5]. A study from the Netherlands reports that 44% of all prescriptions in a paediatric ward are off-label [6]. In Germany, around 40% of under 18s were prescribed at least one off-label medicine among a study of 17 000, with no significant differences according to region, urbanity, migrant background and social class [7].

To summarise these figures, one could say that higher rates are seen in younger patients and in hospital settings, and that a figure of 20% lies at the lower end of these reports. However, consistent estimates of the prevalence of off-label use are made more difficult because they are often not recorded in a patient's notes; this in turn may reflect the fact that they are associated with increased liability for physicians. Thus, it is quite possible for an audit of physician practice to deliver a result indicating a falsely low rate of off-label prescriptions and, where there is a range of figures, to suspect the higher proportion to reflect more accurately the real situation [8].

In fact, in many areas, off-label use is more common than use according to the approved label, bringing to mind the point that in such circumstances the pharmaceutical regulatory system is not fit for purpose. But also, even though this is clearly a very large issue, getting hold of reliable statistics is something of a problem in itself. Off-label medicine is not universally shady, but it does have shady patches, and few practitioners will admit to having participated in the darker regions of the practice any more than they absolutely have to. So there are questions about the statistics, but if they are wrong, one would suspect them to be under- rather than overestimates. Very few doctors would voluntary admit to prescribing off-label when they have not. That also tells you something about the perceived ethics involved. Nevertheless, to avoid criticism, I have erred on the side of caution in my overall statement that it constitutes '20% of all prescriptions'. A widely referenced article looked in detail at the issue and came to a similar conclusion; they also assessed the proportion of off-label use by therapeutic class [9] (Figure 1.1).

To give you some simple examples, the prescription of antibiotics for colds and flu is almost entirely without patient benefit but at significant cost to the NHS in the United Kingdom (and equivalent payers in other countries) and raises concern in an era of increasing bacterial resistance; the prescription of antipsychotics to dementia patients without their consent and at their increased risk is a scandal that led to a recent UK government report and action; and the prescription of antidepressants to children and adolescents when they had only been licensed for adults revealed age-related increases in suicide risk, with increasing risk for young patients but not for old.

Off-label medication is not always a bad thing, and it would be a grave mistake to ban the practice entirely. I certainly would not advocate its prohibition, far from it. In my work on the area of secondary and tertiary uses for existing drugs, I have come to realise the huge potential of this area of study. A main purpose of this book is to ensure that the beneficial discoveries made by doctors in the privacy of their patient consultations are properly validated and widely disseminated. The advantages of this approach are shown clearly by the story that follows, representing one of the

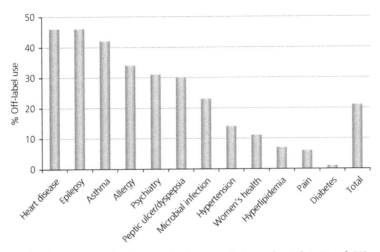

Figure 1.1 Off-label mentions by therapeutic class. Graph drawn from data in Ref. [9].

strangest examples of a bad drug made good through off-label prescription, coupled with a strong element of serendipity.

The drug is thalidomide, a name which connotes some of the worst aspects of pharmaceutical industry misbehaviour and patient harm. Thalidomide was first introduced in 1957 by the West German company Chemie Grünenthal GmbH with the trade name Contergan, a potent and apparently safe sleeping pill. In laboratory rodents, unlike barbiturates – with which it was compared at the time – thalidomide proved remarkably 'safe', insofar as it was almost impossible to administer a single lethal dose. As we know now, these tests were insufficiently broad to cover the full range of toxicological consequences of the drug's long-term administration. Clinical testing in Germany was unsystematic, with pills distributed to employees and samples given to local doctors. With its apparent safety advantages compared to other sleeping pills like barbiturates, which can be lethal at small multiples of their therapeutic dose, thalidomide gained widespread popularity in Europe and Canada; it could even be purchased without a prescription. This was an era of burgeoning use of pharmaceuticals, and their use in psychiatric conditions, as the Rolling Stones recognised so acutely in 'Mother's Little Helper', a song about the widespread use of diazepam (Valium™). It was also an era of minimal regulatory supervision of the pharmaceutical industry. Later on, in addition to its use as a sleeping pill, thalidomide also became popular in the treatment of pregnancy-related morning sickness.

The first 'thalidomide' baby was born on Christmas Day, 1956, before the drug went on the market; she was born with no ears as the daughter of an employee of Chemie Grünenthal who had given his pregnant wife some of the free tablets. Around the same time, physicians and neurologists reported an increased incidence of peripheral neuritis (tingling hands and feet) in adult patients who were taking the sedative. The connection between these cases and the use of thalidomide was not yet clear, but the neuritis effect prevented the approval of the drug by the Food and Drug Administration (FDA) in the United States.

The Australian obstetrician William McBride was instrumental in connecting the use of thalidomide with its toxicity to the unborn child (teratogenicity). He prescribed the drug for women suffering from morning sickness and then suspected a causal link in the malformed babies he delivered months later. McBride led the uncovering of the scandal, which included overcoming the initial intransigence of the drug companies that were involved and the important role of journalists in securing proper compensation for the victims. One estimate is that thalidomide caused malformations in between 8000 and 12 000 infants, 5000 of whom lived to adulthood. In addition, over 40 000 people suffered from peripheral neuritis. Thalidomide was banned by the World Health Organization (WHO) in 1962 and withdrawn from the market in Europe and Canada, and one would have thought that would be the end of its medical life.

But in 1964, a critically ill patient with erythema nodosum leprosum (ENL), a complication of multibacillary leprosy, was referred to Dr. Jacob Sheskin, who was at Hadassah University in Jerusalem by the University of Marseilles, France. The story is brought to life in the eminently readable *Dark Remedy: The Impact of Thalidomide and Its Revival as a Vital Medicine,* by Trent Stephens and Rock Brynner [10]. Leprosy (Hansen's disease) is a chronic, infectious disease of human beings that primarily affects the skin, mucous membranes and nerves. This bacterial disease is caused by *Mycobacterium leprae,* and is normally contracted through the respiratory tract and is similar to the bacillus that causes tuberculosis.

'My discovery', said Dr. Sheskin,

came about only by chance. In 1964, we had a letter from the University Hospital in Marseilles saying that a Moroccan immigrant from the Atlas Mountains on his way to Israel had been hospitalized there with Hansen's disease. They requested permission to allow the patient to continue his travels to Israel [11]. Through the Jewish Agency we answered that since the man was a Jew he was entitled to return to Israel. A separate plane was hired, and the patient, his wife, a son of a previous marriage also suffering from Hansen's disease, and several children of the wife's previous marriage were put on the plane in isolation and sent to Israel.

The patient arrived in a terrible state, suffering from extreme lepra reaction with all the classical symptoms. I have never seen a leper in worse condition. For 19 months he had been bedridden and had not slept for more than 2–3 hours during the course of any 24 hours. The pain he was suffering was unbearable. He had cachexia[3] and was on the verge of death. I had read in the literature that women mental in-patients were being given thalidomide to make them sleep, and that the drug was effective where nothing else had helped. Since the patient was in such a hopeless state, I decided to try thalidomide to sooth him.

After 1 day of the thalidomide – two pills – he slept for about 20 hours without waking. When he woke he said he felt so much better that he wanted to get off the bed to go to the toilet! I only had 20 thalidomide tablets: the drug had already been condemned all over the world and had been withdrawn from the market. I gave him two more pills and made him stay in bed for another 24 hours. This time, the pain, which had been so great, had disappeared almost entirely. After another 3 days of treatment, I decided to stop the pills to see what would happen.

All the symptoms returned, the pain, the insomnia, and the inflammation of the eyes. So I started the pills again. Once again, his condition improved dramatically. I then knew that it was not by chance that he had gotten better but because of the thalidomide. I secured more thalidomide and started to give other patients in the Hospital suffering from extreme lepra reaction the thalidomide treatment. All recovered dramatically.

---

[3]Cachexia, otherwise known as the wasting disease, involves the loss of muscle and other bodily tissues in response to severe chronic conditions, particularly cancer, heart failure and emphysema.

Because of the disastrous effects of thalidomide on foetuses, we did not give the drug to fertile women lepers. We did not want any more deformed thalidomide babies. So we gave the pills only to women rendered infertile. But there are still about 12 million lepers in the world, half of whom are not being treated at all. Much remains to be done to reach them'.

To confirm his findings, Sheskin needed to travel to Venezuela, where there were many more lepers and where thalidomide was still available. There he conducted clinical tests with patients whom he had treated in the past and for whom he had records. He ended up treating 173 patients, and over 90% were symptomatically cured. His unexpected discovery led to further research on the discredited drug; scientists began to ask if it could be useful in other inflammatory conditions? In so doing, they revealed that thalidomide could modify certain immune reactions and could be useful also for those infected with HIV/AIDS and in other autoimmune conditions. More widely, research in cancer revealed that it could inhibit the proliferation of blood vessels associated with the development of tumours, thus slowing tumour growth. This effect paralleled the original effects of the drug on the growing foetus, where it was shown to interfere with the blood supply to the developing limbs of the foetus. We seem to have a common mechanism for the teratogenicity and the oncological activity.

As far as leprosy was concerned, the development was taken up by the US pharmaceutical company Celgene, and after some confirmatory placebo-controlled trials, thalidomide was finally approved in 1997 by the FDA for treatment of ENL. It subsequently became approved for cancer of the bone marrow, or multiple myeloma. To enhance its safe application and use, a special educational booklet was created, called System for Thalidomide Education and Prescribing Safety, to which patients must adhere. As a further indication of its concern about safety and to remind patients of the terrible former history of the new treatment, Celgene wanted to retain the name 'Thalidomide' as a trade name for its product, so that people would recognise its potential problems and avoid them.

This is a remarkable story, elements of which I shall refer to in later parts of this book. Firstly, it is profoundly apposite for my subject matter, since off-label uses refer to the use of medicines outside regulatory purview. Thalidomide was the reason why our current system of medicines regulation exists. Regulatory bodies around the world were drastically strengthened after the tragedy in order to prevent it happening again. A whole range of additional preclinical toxicology tests became necessary before experimental drugs could be first tested on humans. Drug regulation nowadays is all about risk and reward, safety and efficacy. It is a complex process, and enormous amounts of information are accrued and evaluated in order to make the assessment as accurate as possible. However, we recognise that no drug is perfect, and all have side effects. We need to minimise the risk and optimise the reward, but we are never going to have a completely safe or a completely efficacious treatment. It is a balance, infinite shades of grey. The system got it wrong with the first life of thalidomide, but got it right in the second.

So secondly, this story makes another important point. We should recognise that thalidomide's old uses, for inducing sleep and for morning sickness, are troublesome for those affected, but not more than this: they are hardly life-threatening. People who are not treated are likely to be more tired, or nauseous, but these are natural processes which to a certain extent are self-correcting. Patients with insomnia eventually fall asleep; pregnant women with morning sickness eventually move into their third trimester, and the symptoms normally resolve. In comparison, leprosy and multiple myeloma are very serious conditions. They do not self-correct; left untreated, patients with these conditions often die as a result of them. This difference is vital if we are fully to understand and find better systems to deal with off-label uses for existing drugs.

But in another respect, this is a good demonstration of off-label prescription leading to patient benefit, rather than the reverse. So, where then is the harm?

Scope of the issue

Clearly from the thalidomide example, serendipity plays a central role in the history of the development of medicines. Indeed, Dr Sheskin's motivation for prescribing thalidomide to the leprous patient was to enable him to sleep, given that the patient had not been able to do so before coming into his care. Long before the modern era, treatments were often applied to patients in all sorts of conditions in order to discover whether they worked. If you look at the potential uses proposed for herbal medicine, which often derived from folk medicine, you can see the huge range of ailments which they are proposed to work for.

In William Withering's *An Account of the Foxglove*, written in 1785, the author is a practising doctor who tries his treatment on all sorts of patients. Included in the account are examples of diseases we can recognise, like asthma or gout, as well as ones which are difficult to decipher, like dropsy (an old word for oedema, or swelling) or anasarca (which is extreme generalised swelling). These latter two are symptoms of a number of possible illnesses. He informs us in particular that the preparation has beneficial effects on urine production. In days of bladder stones and other urinary problems, the delivery of 'a healthy flow' was generally considered a very good thing. It would particularly be so for patients with extreme general swelling, which could be due to liver failure, kidney failure, heart failure as well as severe malnutrition or protein deficiency. Nowadays, we recognise the active ingredient of the foxglove (*Digitalis purpurea*) to be digoxin, which is used both directly and as its derivative, digitoxin, for the treatment of heart failure. But it is not useful for liver failure, or malnutrition, and when used in excess, digoxin and digitoxin are both poisonous.

The story of *Digitalis* shows how one herbal product was generally used in a number of indications, but over the years, its utility focussed upon one indication. The evidence base closed in around heart failure. Dr Withering, in the late eighteenth century, played his part in this process by providing a written account of his experiences. Nowadays, drugs are subjected to randomised, controlled trials to find out whether they actually do any good and then first introduced for one indication alone, to be given to a specified patient population, at a specified dose and in a specified form. Subsequent to that introduction, experimental efforts are made to expand the limits of these specifications. Sadly, unlike Withering's treatise, not all of the experiences in that process of expansion are documented in modern medical practice. Generally, the prescription of a drug in an off-label fashion, based as it is on limited evidence, does not itself result in the enlargement of the evidence base; the result of the prescription is not written and disseminated for future cohorts exposed to similar treatment. This problem is further dealt with in Chapter 8, where I discuss whether it is right that off-label medication should be categorised as different from a clinical trial, as it currently is, and how we can routinely document off-label uses, so that we can build the evidence base for safety and efficacy upon which future prescribers can depend.

One of the major classes of off-label drug use is for paediatric therapy. For legal, ethical or practical reasons, clinical trials are usually not performed on children (nor are they routinely conducted on other patient groups, such as pregnant women or senior citizens). In paediatric respiratory care, many drugs are not available in formulations suitable for infants and toddlers, having been tested predominantly on older children. However, respiratory drugs are frequently used in children for common diseases like asthma, upper and lower respiratory tract infections, rhinitis (allergies) and sinusitis. Three-quarters of marketed prescription drugs have no labelling indications for children, although their inclusion in clinical trials is enlarging [12]. Among medicines which were newly licensed by the European Medicines Evaluation Agency (EMA) between 1995 and 2005, only one-third was specifically licensed for children [7]. Thus, off-label use is particularly widespread in paediatric situations, and over half of children in Europe who are prescribed medicines in hospital receive a medication that is either 'unlicensed' or 'off-label' [13]. Other studies put the figure at 40%, 45% or 76% [6,7,14], and the area has been comprehensively reviewed [15]. To summarise these statistics, the younger the patient and the more critical and rare their illness, the more likely they will be treated off-label. This may be because the licence is for older children, whereas the prescription is for a younger child. It may also be for a different use than that on the approved label; or it may be because the dose is different or its schedule of administration is different from the approved use. Commonly, it is believed the problem is that the drug is approved for adults, having been tested only in adults. There are regulatory incentives available to producers through the regulatory system, and the extent to which this can be improved is dealt with more thoroughly in Chapter 7.

So what? Do these statistics amount to anything of real patient import, or am I just demanding unrealistic standards of pointless box-ticking by our medical practitioners? Of course, if there were no adverse consequences, there would be little to complain about (apart from the waste of money). So, for one thing, let me point out that there is an increased rate of adverse drug reactions (ADRs) associated with off-label use. A study from France in paediatric care showed that there were over three times as many ADRs associated with off-label medication as with 'on-label' alternatives [16]. The relationship does not just apply for ADRs overall; it is particularly concerning that a Swedish study showed that off-label drug use in children was associated with a significantly higher number of *serious* adverse reactions [17]. A study from Liverpool, United Kingdom, also showed that the rate of adverse reactions associated with off-label medication, at 6.0%, was significantly higher than on-label comparators, at 3.9% [18]. Another report from Derbyshire, United Kingdom, cited two studies, one of which suggested that five out of eight severe ADRs were associated with the off-label use of a medicine. This study suggested that the percentage of unlicensed and off-label drug use was significantly associated with the risk of an ADR. The other study found that 14 of 19 drug prescriptions associated with 17 severe ADRs were either unlicensed or off-label [19]. In a study of ADRs in children and adolescents over a 10-year period in Denmark, 60% of ADRs reported for medicines prescribed off-label were serious, and, in contrast, 35% of ADRs reported for 'on-label' medicine use were serious. Thirteen of the off-label serious ADRs resulted in a fatality [20].

So, safety in paediatric medicine is something that we will need to pay close attention to as we go forwards in our consideration of patient consent and the ethics of off-label medication (see Chapter 3), especially as safety is a paramount consideration when children are concerned, and consent often has to be given by the patient's parents, most of whom are acutely sensitive to safety concerns.

Psychiatric care is another common area of off-label use. A US study looked at psychotropic drugs over a decade-long period from 1998 through 2009: the average proportion of all uses that occurred off-label was 23.3% for antidepressants, 60.7%

for antipsychotics and 54.2% for mood stabilisers [21]. At these high levels, there are more drugs being prescribed off-label than in accordance with the regulatory approval; in this situation, this 'is the death of regulation' [3]. But it gets worse, as we now see.

In November 2009, the UK Department of Health issued a report on the prescribing of antipsychotic drugs to people with dementia [22]. With a rare exception, no antipsychotic is licensed in the United Kingdom for treating the behavioural and psychological symptoms of dementia; the approved label is generally limited to the treatment of schizophrenia and bipolar disorder. This therefore constitutes off-label medication except in the rare and unfortunate event that the demented patient also has a co-morbid psychotic illness. The report had been commissioned in recognition of widespread concern about the overprescription of antipsychotic drugs to patients with dementia, a concern felt particularly among patients and their families who had not been consulted before such antipsychotic treatment began; neither were they aware that the prescription was being administered in an off-label fashion. Subsequently, the UK National Institute for Health and Care Excellence (NICE) conducted its own evaluation [23]. It was suggested that up to a quarter of people with dementia had been prescribed antipsychotics in addition to their normal medication, up to 180 000 people at a cost of £90 million a year. In formal care, the proportion of dementia patients being prescribed antipsychotics rose to nearly a half [24]. The vast majority of these, 8 out of 10, derived no benefit from antipsychotic treatment, and NICE advised against the use of any antipsychotics for non-cognitive symptoms or challenging behaviour of dementia unless the person is severely distressed or there is an immediate risk of harm to them or others.

Moreover, not only are these drugs ineffective, but they are dangerous in older patients. There is an approximately threefold increased risk of cerebrovascular adverse reactions (in other words, stroke) that have been seen in randomised placebo-controlled clinical trials in the dementia population with some atypical antipsychotics, including risperidone, aripiprazole and olanzapine. As a result, NICE worked out that for each 100 patients treated with atypical antipsychotics over a year, one of them died prematurely and a similar additional number suffered non-fatal cerebrovascular events, such as stroke (around half of which may be severe) as a result of adverse side effects. In the context of longer-term treatment, there were estimated to be 167 additional deaths among every 1000 people with dementia treated with antipsychotics over a 2-year period. A similar situation pertained in the United States, where the Department of Health and Human Services found that in a 6-month period in 2007, 14% of nursing home residents were given antipsychotics, presumably with similar adverse outcomes [25]. Of course, stroke is not the only side effect of antipsychotic drugs: they also produce over-sedation, which may result in falls or in patients to become 'like zombies' [24], and worsening of cognitive function, which is exactly the opposite of what medical treatment of dementia sufferers is supposed to achieve.

Antipsychotic prescribing is also a surprisingly common practice for active duty troops in the US military, even though psychotic illness itself would presumably not be desirable, and should have been screened out as part of the selection process. Nevertheless, antipsychotic prescription is a growing trend: it was reported [26] there has been a 682% increase in the number of psychoactive drugs – antipsychotics, sedatives, stimulants and mood stabilisers – prescribed to US troops between 2005 and 2011, despite a steady reduction in combat troop levels since 2008. Some of the off-label uses for which these drugs have been prescribed include insomnia, anxiety and aggressive behaviour. They have also been used to treat post-traumatic stress disorder (PTSD), even though the evidence for efficacy in this area is weak (for instance, one antipsychotic risperidone failed in a clinical trial examining its use as add-on treatment of PTSD symptoms [27]). Certainly, by comparison with antidepressants, which are demonstrably effective in PTSD, antipsychotics are very much the inferior option. In addition to the nearly sevenfold increase in the use of antipsychotic drugs, there has been an even bigger increase in the use of sedating anticonvulsants, an increase that is not matched in the general population, and, perversely, a decrease in the use of antidepressants. Given the comments earlier about PTSD and depression, which are both likely outcomes of active duty, the decline in antidepressant use is disturbing if they are being substituted by less efficacious antipsychotic alternatives, with consequently poorer therapeutic outcome. It is particularly disturbing given that untreated PTSD and depression are both associated with increased rates of suicide, and in soldiers involved in the Iraq war and Afghanistan mission, suicide is a greater risk than death in combat.

The overall prevalence of off-label prescribing in psychiatry is high. A study from 2000 reported that 65% of psychiatrist respondents had prescribed medication off-label within the past month [28]. A German prescription survey from 2000 reported that antipsychotic drugs were being prescribed in older patients (aged 49–70 years) almost exclusively for off-label indications [29]. Another study found the off-label use of Eli Lilly's atypical antipsychotic Zyprexa (olanzapine) to be 45% [30]. However, one of the problems in psychiatry is the complexity and specificity of the diagnosis, together with an increasing specificity of the label. It may be helpful to think of a spectrum of use of licensed psychotropic drugs in unlicensed applications, with some off-label prescribing being nearer the label than others. Because of this very fine level of diagnostic specificity, there may be no regulatorily approved treatment for the specific condition to be treated [31]. To address this problem, I came across a blog post in the course of researching this book [32] advocating that psychiatrists should formally diagnose in accordance with the approved label in order to avoid their prescription being regarded as 'off-label', whatever the real patient diagnosis. The [anonymous] author of this blog appears to be a pharmaceutical industry insider or a medical professional. It is not clear whether this is a serious proposal or to what extent if reflects widespread, if unspoken, attitudes among medical professionals for their patients. However, other reports have used the term 'diagnosis shifting', to indicate physician behaviour in cases where they are aware that prescribing practices are being monitored. For example, when treating a patient with sinus congestion but prescribing antibiotics, a clinician might be more inclined to write it up as sinusitis, an antibiotic-appropriate diagnosis, instead of nonspecific upper respiratory infection, an antibiotic-inappropriate diagnosis [33].

Part of the problem is lack of awareness among prescribers about exactly what drugs are approved for what indications. A US survey of nearly 1200 physicians (599 primary care physicians and 600 psychiatrists) showed widespread ignorance of what drugs were actually approved for. The study was conducted in 2007–2008 and included 22 drug–indication pairs. The indications varied in their FDA approval status from on-label use, to off-label use supported by medical evidence, to off-label use deemed to be ineffective. In the area of psychiatric care, 13% of all physician respondents erroneously believed that quetiapine (Seroquel®) was FDA approved for dementia with agitation; an even higher number, 19%, had prescribed quetiapine for dementia with agitation, suggesting that roughly one-third of doctors had prescribed this indication pair knowingly off-label, with roughly two-thirds having prescribed it unknowingly [34]. These two groups may be categorised, somewhat unkindly, but not inaccurately, as the 'ignorants', those that do not know, and the 'insouciants', those that do not care. In this case, it is difficult to know which group to be most concerned about.

A common form of off-label medication in psychiatry is the prescription of drugs at higher than their approved dosage ranges. While potentially offering greater efficacy, for instance, in patients not well treated for depression, the likely adverse consequences of this practice on safety are obvious. Another factor which seems to be more prevalent in psychiatry is that the approved indications change over time, with some new uses being added to the list of approved indications (such as the expanding uses beyond depression for serotonin uptake inhibitors, [SSRIs] for anxiety and obsessive compulsive disorder), while sometimes the approved uses can be reduced (for instance, the use of fluoxetine for depression in children and adolescents) [35]. So, we need to reflect that the situation is complex and nuanced, but that in itself is not a good reason for an abandonment of concern over the scope of off-label medication in psychiatry [36].

When we put paediatric and psychiatric use together, as in paediatric psychiatry, off-label prescribing is even more common, and other concerning facets arise. In the United States, it has also been estimated by the American Medical Association that although certain atypical antipsychotic drugs are FDA approved for specific uses in paediatric patients, the majority of prescribing (70–75%) is off-label for these drugs in this patient group.

One of the additional concerns with atypical antipsychotics involves their adverse metabolic effects. Because the risk of childhood obesity is inversely related to socioeconomic status, low-income children who are already at high risk for obesity and related metabolic disorders may be especially vulnerable to the adverse effects of weight gain from atypical antipsychotics. Children treated within the Medicaid programme, that is, publicly funded healthcare for the poor, were four times more likely to receive antipsychotics than children treated privately [37,38].

A substantial proportion of paediatric psychiatric prescriptions are written to control violent behaviour, rather than address the root cause of the violence, which would require more expensive psychotherapeutic intervention. The inference is that the expensive, non-medicated option is only available for private (non-Medicaid) care, and this explains the fourfold lower rate of antipsychotic prescription in these patients. Some of the prescriptions were written for very young children: for example, some children between ages 1 and 2 received antipsychotics for conditions such as autistic disorder and attention deficit disorder (ADD) with hyperactivity. In this context, we should note that between 1994 and 2003, reported diagnoses of paediatric bipolar disorder increased 40-fold, from about 20 000 to approximately 800 000. That diagnosis was associated with the claim that extreme irritability, inattention and mood swings were actually due to paediatric bipolar disorder that can occur before age 2. This increase in incidence seems difficult to reconcile with a real change in psychiatric disorder, but of course, once a diagnosis associated with bipolar disorder is reached, it is much easier to prescribe an antipsychotic, albeit one that is not specifically approved for paediatric use. In 2008, an estimated $6 billion was spent on off-label antipsychotics in the United States, of which $5.4 billion was for uses based on uncertain evidence [39].

The US Department of Health and Human Services recently reviewed antipsychotic drug use by Medicaid recipients of age 17 and under, focussing in particular on the newer 'atypical' antipsychotics, which include AstraZeneca's Seroquel (quetiapine), Eli Lilly's Zyprexa (olanzapine), Johnson & Johnson's Risperdal (risperidone) and Otsuka Pharmaceutical's Abilify (aripiprazole). Medicaid spends more on antipsychotics than on any other class of drugs, and there is evidence that 70% of the cost of these drugs in the United States was paid for by Medicaid and other government programmes. In 2013, aripiprazole became the top-selling drug in the United States, reaping revenues of $6.3 billion for the Japanese pharmaceutical company Otsuka in that year.

The last general area where substantial off-label use exists is in oncology. The levels in this area has been estimated at between a third [8] and a half [40], with another study suggesting that 50–75% of all uses of drugs in cancer care in the United States are off-label [41]. In terms of cost, in 2010, $4.5 billion was spent on off-label chemotherapies compared to an overall bill of $12 billion. It has been estimated that 62% of cancer patients use drugs off-label. In cancer, although regulatory approval is generally given for a specific type of cancer, it is a reasonable and frequently valid hypothesis that efficacy crosses into other types. Moreover, given the seriousness of the condition, there is much public sympathy for the widest possible armoury of therapies to be available to the patient, regardless of actual regulatory status. The European Society for Medical Oncology, for instance, is dedicated 'to promote equal access to optimal cancer care of all cancer patients' in its mission statement. In reflection of this sentiment in the United States, in 2008, Medicare rules were changed to cover more off-label uses of cancer treatment drugs (the issue of drug reimbursement is dealt with fully in Chapter 6). A 2008 study found that 8 out of 10 cancer doctors surveyed had used drugs off-label. Studies have reported that about half of the chemotherapy drugs used are given for conditions not listed on the approved drug label. Nevertheless, the evidentiary support for such off-label use is very variable in quality, and always less than would be required for a formal regulatory approval [42].

In most cases (not just in cancer), adequate research evidence to support off-label prescribing is lacking. A survey of 150 million off-label prescriptions in the United States found that 73% had little or no scientific support, even when sources other than the product information were searched [9]. A similar percentage, 79%, of off-label medicines were found to lack scientific support from a survey in Canada [43]. Similarly, in Australia where the prevalence of off-label prescribing overall is between a wide range of 7.5% and 40% in adults and may be up to 90% in some hospitalised paediatric patients, there is a lack of evidence to support prescriber's off-label choices [44]. As a consequence, only a minority of off-label prescribing can be said to be scientifically justified. When analysed by therapeutic area, a similar if not more alarming situation pertains (see Figure 1.2). In terms of cost, Stafford et al. looked at the expenditures on off-label antipsychotics in the United States over the period to 2008 and found a huge increase in poorly evidenced use that reached $5.5 billion a year by the end of the period, relative to $4 billion of on-label use and around $0.7 billion of well-supported off-label use (see Figure 1.3) [45]. This matter is further dealt with in Chapter 4.

Aside from the general therapeutic areas, there are some specific cases where off-label rates are shockingly high. A particular example is NovoSeven, marketed by Novo Nordisk A/S of Denmark; it is a bioengineered form of factor VIIa, a critical protein involved in the coagulation of blood [46]. Factor VIIa is approved by the FDA for patients with certain forms of haemophilia or congenital deficiencies in the protein, for whom it can prevent potentially fatal episodes of bleeding. But an anecdotal story emerged involving a soldier that had received life-saving treatment of a battlefield wound. In June 1999, an Israeli doctor, Uri Martinowitz, injected two doses of factor VIIa into a 19-year-old Israeli soldier who was shot through the abdomen with a high-powered rifle; it was the first reported use of the drug in a non-haemophiliac

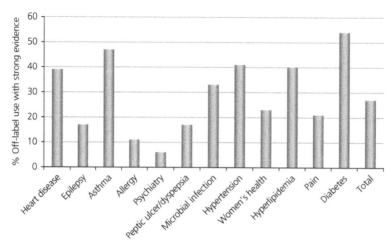

Figure 1.2 Levels of off-label use and scientific evidence, by therapeutic area. Graph drawn from data in Ref. [9].

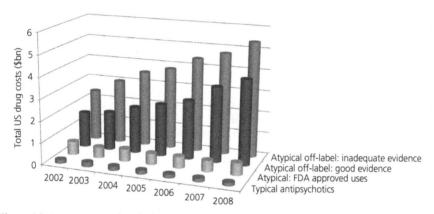

Figure 1.3 Costs associated with the prescribing of antipsychotic medications in the United States, 2002 through 2008, categorised by off-label status and level of supporting evidence. Reprinted with permission from Ref. [45]. © Nature Publishing Group/Macmillan. (*See insert for color representation of the figure.*)

and, according to Martinowitz, caused the bleeding to 'stop immediately'. Between 2000 and 2008, industry-sponsored research revealed that factor VIIa could also prevent bleeding in heart surgery, trauma or haemorrhagic (i.e. bleeding) strokes. As a consequence, annual usage soared 140-fold (while use in haemophilia increased less than fourfold), from about 125 to a total of more than 18 000 cases; over the total 8-year period, it is estimated that factor VIIa was used in more than 70 000 hospital cases in the United States [46].

By 2008, the latest year studied, researchers found that off-label uses accounted for 97% of orders for the drug, total global sales of which amounted to $1.43 billion. The figure reflects the high cost of the drug, which can result in hospital costs of up to $650 000 for a single patient at the highest doses [47]. Part of the reason for the high cost is the way it is manufactured, using biotechnology rather than large-scale chemical reactors, but drug costs also increased because recommended dosages increased. Factor VIIa is a complex biological protein that is made in hamster kidney cells that have been genetically engineered to secrete factor VIIa; the protein is harvested outside the cell and purified from the extracellular serum. The cost of factor VIIa, though high, is similar to other biological drugs, which are often produced by expensive manufacturing processes.

So far, then, the argument is that this drug is hugely costly and widely sold by means that make a mockery of the drug regulatory system. But a further key question is, 'Does its off-label use nevertheless benefit patients?' The clinical evidence in favour of most of the off-label use of NovoSeven is weak and disrespected. In 2011, researchers from Stanford University in California published a systematic review of the benefits and harms of factor VIIa when used for five off-label, in-hospital indications [48]. These were cardiac surgery, intra-cerebral haemorrhage, liver transplantation, prostatectomy (prostate removal) and trauma. They identified all the available evidence that addressed this issue and found 64 reports worthy of review. Only 16 were randomised, controlled trials; the others were observational studies that compared the treatment with another approach or had no comparator group. The authors concluded that limited available evidence for these five off-label indications suggested no mortality reduction with factor VIIa use. An editorial that accompanied this article stated that overall, there was no evidence that factor VIIa reduced mortality for any off-label use; however, it did increase the risk for thromboembolism, in other words a blood clot resulting in a heart attack or the kind of stroke where blood flow to the brain is interrupted. In plain English, off-label factor VIIa does not work and caused harm to patients when so used. Their findings are compatible with other recent studies. The editorialists noted that 'allowing physician autonomy to choose medications is appealing, but not when it results in unhelpful, dangerous, and costly decisions' [49]. This is an example where for every three prescriptions written in 2008 in accordance with the regulatory approval generated from the producer Novo Nordisk's submitted data, the company sold 97 off-label doses. That's a pretty interesting ratio, particularly if, as stated earlier, these 97 doses caused no benefit and harmed patients, but still cost thousands of dollars each.

A second example is bone morphogenetic protein 2, or BMP-2. This product has been approved for orthopaedic usage in the United States and Europe in a medical device, rather than as a drug *per se*. Bone morphogenetic proteins are growth factors that promote bone formation. BMP-2 is a particularly powerful bone growth stimulator, but more than that is thought possibly to be also involved in all kinds of biological processes involving changes in bodily form and structure. There are many roles for dysregulated BMP signalling in pathological processes, including cancer; over-activation of BMP signalling is instrumental in the development of adenocarcinoma in parts of the gastrointestinal tract. Medtronic is the producer of a medical device called InFuse, composed of a collagen sponge containing BMP-2 for the treatment of certain types of spinal fusion procedures in adults with degenerative disc disease, and at one time achieved close to $1 billion in annual sales.

Originally approved in 2002, a series of complications and side effects have grown in magnitude since that time. Many of them occurred when BMP-2 was used in an off-label manner, which was estimated to be about 85% of the time [50]. Medtronic found in a clinical trial of BMP-2 in a type of spinal fusion surgery that there were troubling findings of bone formation in the spinal canal (i.e. outside the area of fusion) of 75% of the BMP-2 patients. The trial was halted, but the results were not published until 2004, 5 years after the trial was conducted. Other cases of ectopic bone formation[4] were observed in a trial of cervical spine fusion. In some cases of cervical spinal fusion, there have been alarming reports of head and neck swelling, resulting in the compression of the airway and neurological structures of the neck. In other reports, there is an increased level of cancer reported in patients treated with BMP-2, particularly when administered in higher doses, and debate as to whether this is causally related.

In 2013, the US Senate Finance Committee published a report on Medtronic's product [51]. The part relating to off-label uses of InFuse reads:

> The FDA's 2002 approval of InFuse was limited to spinal surgeries using the anterior lumbar inter body fusion (ALIF) technique...The Agency for Healthcare Research and Quality (AHRQ) estimates that...'at least 85% of InFuse use is now off-label.' In 2008, the FDA published a public health notification linking the off-label use of InFuse in the cervical spine with life-threatening swelling in patient's [sic] throats and necks. The Wall Street Journal reported at the time that 'the agency... received 38 reports over four years of side effects, mainly swelling of neck and throat tissue, which resulted in compression of the airway and other structures in the neck.' In addition, the Wall Street Journal reported that '[a]t least three-quarters of the roughly 200 "adverse events" reported to the FDA involve off-label uses of InFuse'.

The risk and benefit of BMP-2 has been debated for over a decade. In what is probably a defining moment for the product, a report published by the widely respected Yale University Open Data Access Project in June 2013 in the *Annals of Internal Medicine* detailed the results from two independent groups that 'rhBMP-2 provided little or no benefit compared to bone graft and may be associated with more harms, possibly including cancer...' [52]. The criticism applied to both off-label and on-label uses. Meanwhile, evidence continues to mount regarding the increasing adverse events attributed to utilising BMP-2 'off-label' in spinal surgery [53]. It is a pity therefore that the off-label expansion of the approved use exposed approximately six to seven times as many patients to harm while not offering treatment proven to be effective.

The third example is that of niacin (vitamin B_3), which is used for its 'good' cholesterol-raising properties. In a long-acting form as Niaspan™, it is indicated to reduce elevated total cholesterol and LDL ('bad') cholesterol and to increase HDL ('good') cholesterol. Whereas the mainstay of cholesterol therapy is based on the administration of the statin class of drugs, which have been shown both to control cholesterol levels and to reduce heart attacks and strokes in people with abnormal cholesterol levels, Niaspan and other drugs based on niacin have not demonstrated this functional benefit. In other words, they have only demonstrated an effect in reducing 'bad' cholesterol, not on the consequences of so doing. In 2012, 5.3 million prescriptions costing $1.1 billion were written for Niaspan, according to data provided by IMS Health, on the assumption of such a benefit. James Stein, a professor of cardiovascular medicine at the University of Wisconsin School of Medicine and Public Health, estimated that only about 25% of Niaspan's use was 'proper'; by inference, the other $825 million spent was for 'improper' use of the drug [54].

However, in 2013, the results were reported of a large randomised trial testing the use of extended-release niacin for the reduction of major vascular events. It was called the Heart Protection Study 2—Treatment of HDL to Reduce the Incidence of Vascular Events (HPS2-THRIVE) study. The study found that not only did this special

---

[4]Ectopic bone formation relates to bone being formed in an unwanted place, where it does not belong.

extended-release niacin formation not reduce heart attacks and strokes, it increased the risk of bleeding, infection and diabetes. This example indicates that sometimes the off-label use of a drug is a result of an improper inference based on a biomarker; in the case of niacin, it was assumed that the effects on cholesterol were functionally important in heart disease, but when this was tested, they were not. As in the previous example, there have been accusations of improper marketing by the pharmaceutical company (Abbott); but at the end of the day, it is the prescriber who has assumed, in this case falsely, that the effects on cholesterol would equate to a beneficial cardiovascular outcome, in advance of that outcome having been demonstrated.

Finally, we should recall the interesting story of modafinil. Modafinil was originally marketed in Europe by the small French pharmaceutical company Lafon for idiopathic hypersomnia (in other words excess sleepiness for unknown reasons); but in 1986, the US company Cephalon leased the US rights to the compound and obtained orphan drug designation and subsequent FDA approval to market the compound for narcolepsy, under the brand name Provigil. Orphan drug designation is a scheme put in place by many of the regulatory agencies worldwide to offer market inducements for rare diseases. We will talk about it more in Chapter 4. Narcolepsy is a chronic brain disorder that involves poor control of sleep–wake cycles. People with narcolepsy experience periods of extreme daytime sleepiness and sudden, irresistible bouts of sleep that can strike at any time. These sleep attacks usually last a few seconds to several minutes. Narcolepsy often occurs with cataplexy, a sudden loss of voluntary muscle tone while awake, that makes a person go limp or unable to move. Narcolepsy with cataplexy is estimated to affect about 1 in every 3000 Americans. Given the infrequency of the marketed condition, Cephalon led an astute campaign of increasingly wide regulatory approvals that established modafinil as a 'wakefulness promoting agent' rather than a classic amphetamine-like stimulant, and with those approvals came substantial commercial success, the product selling $744 million in the United States in 2007.

On the basis of this, Cephalon purchased Lafon, the originator of the product, and continued to seek further approvals for additional uses of the compound from the FDA. In the meanwhile, modafinil developed a wide range of additional off-label uses, so much so that its use grew almost 10-fold between 2002 and 2009. During the period from 2001 to 2006, Cephalon was specifically alleged to have illegally promoted modafinil for non-approved uses, for the treatment of sleepiness, tiredness, decreased activity, lack of energy and fatigue. Modafinil also became widely used for ADD (ADHD) and for non-therapeutic uses such as a cognitive enhancer to boost exam performance; it is a drug of choice for the competitive student around revision time or for the workaholic executive who needs to boost alertness after marathon sessions of work and travel. People who take it say it keeps them awake for hours or even days, maintaining most users in a refreshed and alert state but still able to go to bed when they are ready. It is widely available through Internet pharmacies. According to Barbara Sahakian, professor of psychiatry at the University of Cambridge, there is evidence that a staggering 90% of modafinil's use is off-label [55]. The success of modafinil does not result solely from off-label uses, since Cephalon also filed for additional approved uses in order to expand the market for their product; and while much of the recent uptake of modafinil is a result of off-label use, Cephalon currently derives little benefit from this since the patents have expired and the product is mainly sold by multiple generic manufacturers.

Taking these individual examples together, in conjunction with the therapeutic area-wide assessments, it is clear that off-label use is a widespread practice, but with particularly common occurrence in specific hotspots. While prescribers sometimes need to be able to find solutions for rare diagnoses that are not covered by the approved use of any individual drug, the practice of off-label medication goes much wider than that. There are concerns this raises both for the real level of efficacy of a treatment outside its regulatory approval, as well as the safety in the unapproved use or unapproved patient population. It crosses into dubious grey areas where pharmaceutical companies have been found guilty of improper marketing and, as we shall discover in Chapter 7, eye-watering fines. We will need to ask whether, and under what circumstances, reimbursement of prescription costs to the patient in respect of these off-label uses is appropriate. Having sampled the scope of the issue, we can also delve more deeply into these safety and efficacy aspects. But before we do, let's understand why this practice has grown up, because there is a legitimate basis for finding drugs to work in more than one therapeutic use. Then once we have looked at some of these examples, we can ask, as I have, whether and how the current extent of off-label use can be justified.

CHAPTER 2

Where it all went right: new uses for existing drugs supported by good evidence

In the seventeenth century, the philosopher Francis Bacon cited three inventions which had transformed his world. They were printing, gunpowder and the compass, and they all emanated from China but in very different forms to their original conception. The compass was invented as a child's toy, gunpowder was for incorporation in fireworks, and printing was a technique to preserve ancient Buddhist texts. But by the time they reached Europe, each had found a much more historically important use. The compass became a means of finding new trade routes and destinations, printing was a tool for religious sedition in the Reformation, and gunpowder in enormous quantities was the principal component in the eponymous plot which nearly destroyed the English seat of parliament in 1603. Clearly, it is not just in medicine that the presentation and the use of an invention in practice can be quite different from the original intention.

What follows is a somewhat random collection of vignettes involving secondary uses for existing drugs. There are plenty more that I could have included, but this group provides an idea of the breadth of the possibilities. As you will see, the selection covers a wide range of therapeutic areas, both for the original use and for the secondary use. However, what binds them together is that the secondary uses are all supported by substantial quantities of clinical evidence. In the first group of cases, they have been through regulatory scrutiny and received approval for the secondary use. This gives us confidence that the safety and efficacy of the products is appropriate. The second group then gives a couple of examples where there is strong evidence but regulatory approval has nevertheless not been obtained.

Examples where products have been through regulatory approval for a secondary use

Finasteride: pseudohermaphroditism and hair growth

Finasteride is a drug that was invented both for its original and its secondary use by a large US company, in this case Merck. The story began in 1955 with the finding by two successive anthropologists of pseudohermaphroditism due to in-breeding. Hermaphroditism means having reproductive organs normally associated with being both male and female (see Figure 2.1) [56]. In this case, the afflicted were men thought to have been born as girls, but they really were males, so it was called 'pseudohermaphroditism'. They had a genetic disorder due to in-breeding; their discovery arose from studies of 24 affected subjects from 13 families in a large group from the Dominican Republic and in two siblings from Dallas, Texas. In some cases, the parents and grandparents on one side were first cousins. Since the original discoveries,

Off-label Prescribing – Justifying Unapproved Medicine, First Edition. David Cavalla.
© 2015 John Wiley & Sons, Ltd. Published 2015 by John Wiley & Sons, Ltd.

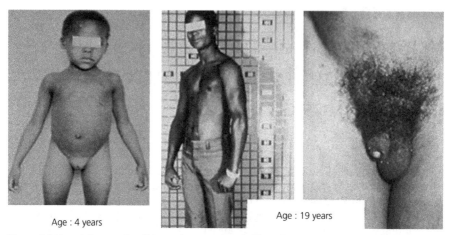

Age : 4 years   Age : 19 years

Figure 2.1 Pseudohermaphroditism; photographs of afflicted males before and after puberty. Reprinted with permission from Ref. [56]. © Elsevier.

other groups have been found, in New Guinea and even more recently in Turkey. As these boys grew into puberty, it became clear that they had small penises, scanty beard growth, no hairline recession, no acne and small prostates. In other words, their genotype had resulted in certain differences which many men would consider advantageous, like no hairline recession. Baldness can be considered a cosmetic issue which makes men look older; but another of the physical differences in the list, prostate enlargement, can give rise to urological problems of a more serious nature.

Merck took up the anthropological finding as the starting point for a drug discovery programme guided towards the more serious medical issue, prostate enlargement. The phenotype of these pseudohermaphrodites was found to be due to a deficiency of an enzyme along the testosterone pathway, 5-alpha-reductase, which is responsible for the final stage of testosterone production, from dihydrotestosterone (DHT). Merck hoped that if they could produce a similar biochemical consequence in men with large prostates, they might prevent prostate enlargement. Both testosterone and DHT are necessary for complete male sexual differentiation and development, but for certain male characteristics, DHT alone can suffice. Blocking the final stage conversion of DHT to testosterone has no adverse outcome on certain male attributes. Thus, the lack of the 5-alpha-reductase enzyme was not associated with lack of libido, height or musculature; affected men even had deep voices. The selectivity of action of the 5-alpha-reductase enzyme was positive for the development of drugs to affect the undesirable actions of testosterone without impairing the desired masculinity of potential patients.

This biochemical understanding took 20 years from the original anthropological report. Merck then began the drug discovery process to find an inhibitor of 5-alpha-reductase, which took another 20 years before their first product, based on the drug finasteride, was launched in 1992; the medical indication was benign prostatic hypertrophy, and the product was brand named Proscar™. Pharmaceutical R&D is a long process; by the time the first product was launched, the anthropologist behind the original discovery was dead.

Subsequently, Merck undertook to develop a product for the treatment of the less serious medical condition associated with the genetic deficiency, male pattern baldness. This time, the product development was much quicker, since the active ingredient was already known and some of the information from its primary development campaign could be reused. They took finasteride and reformulated it into a fivefold lower-dose tablet, called Propecia™, and launched in 1997. Although based on the same ingredient, they were able to file another patent for its use in this particular condition. Propecia was as a result patented until 2013, when generic competitors came on the market.

Hair loss is an area where another efficacious product, this time a topical one, also originally started life in another therapeutic area. Minoxidil was originally introduced as an oral treatment for severe hypertension under the brand name Loniten™. However, after clinical use began, it was accidentally discovered to have an interesting side effect, to stimulate hair growth in men. Its mechanism of action appears to be as a potassium ion channel opener, dilating the peripheral vasculature and increasing peripheral blood flow. It has a direct effect on the follicles, increasing proliferation and differentiation of epithelial cells in the hair shaft. The oral product has a range of problematic side effects, reserving itself for serious cases of hypertension that are not amenable to other agents. However, hair regrowth is treated with the topical agent Rogaine™ (Regaine in the United Kingdom), which minimised systemic exposure.

Finally, the most bizarre story of a hair-stimulating effect, this time for women, came from that of an anti-glaucoma agent, bimatoprost. It was approved in 2001 as Lumigan™, an ophthalmic solution; but similar products like latanoprost were accidentally found to increase the diameter, density and length of eyelashes in patients who used them for treatment of glaucoma. Allergan, who made bimatoprost, undertook the formal studies to show that their product worked as an eyelash-growth stimulator. Mechanistically, bimatoprost and latanoprost act to stimulate the prostanoid FP receptor; in glaucoma, this increases the outflow of fluid from inside the eye to lower the eye pressure; in hair follicles, this stimulates the transfer from the telogen (hair falling out) to anagen (hair growing) phase. While there is a medical term for poor eyelash growth, hypotrichosis, the predominant utility for this discovery was cosmetic, and it was soon picked up by that industry. There are now a number of mascara products containing this pharmaceutical. In 2008, FDA approved bimatoprost (trade named Latisse™) for the cosmetic use of darkening and lengthening eyelashes.

Sildenafil: re-tasking the blue pill for a life-threatening illness

The development of sildenafil began in 1986 with chemists at Pfizer searching for a compound to treat hypertension through the inhibition of the phosphodiesterase type 5 (PDE 5) enzyme. PDE 5 is a subtype of phosphodiesterase, an enzyme involved in the breakdown of cyclic adenosine monophosphate (cAMP), an intracellular messenger with multiple properties, and inhibiting the enzyme increases levels of cAMP, with various consequential effects. Test compounds which were shown to inhibit PDE 5 activity, resulted in vasodilatation and platelet inhibition, and the project was shifted towards the therapy of angina, since the cardiovascular effects of PDE 5 inhibition were thought to stimulate blood supply to the heart (angina involves a deficit in this respect). A lead compound called sildenafil was identified, and human trials began in Wales. These studies were disappointing for their primary endpoint, but some patients reported to their surprise the unexpected side effect of penile erections, leading to the development of sildenafil as a treatment for erectile dysfunction, brand named Viagra. This is a well-known, slightly comic story, of how a straightforward cardiovascular programme can lead to a product for enhancement of sexual gratification. However, Viagra's discovery also led to a re-evaluation of the condition of erectile dysfunction, which had previously been understood to be mostly a quality of life issue. After the drug was marketed, the condition became far more openly discussed; for many, particularly those with a co-morbid illness, such as type II diabetes, Viagra was actually a very useful product. The little diamond-shaped blue pill is still used recreationally, but for real patients, it has a serious utility. However, in comparison, the second therapeutic incarnation for sildenafil is for a disease that is much more than a quality of life issue and has life-threatening consequences.

Some time after Viagra was approved for erectile dysfunction, the role of PDE 5 in pulmonary hypertension became better understood, and the use of sildenafil in its treatment was postulated. Pulmonary hypertension is a general term for a disease process resulting in a progressive increase in pressure in the vessels supplying the lungs, particularly the pulmonary artery. It is different from, and much more serious than, general hypertension. It can be idiopathic (i.e. without known cause), familial (i.e. genetic) or secondary to a variety of conditions such as rheumatoid arthritis or HIV. Patients with the disease often present with signs and symptoms of right heart failure, including shortness of breath, dizziness, fainting, leg swelling and so on. It is a life-threatening disease, unlike erectile dysfunction, and if left untreated can be fatal with a median survival of 2–3 years from the time of diagnosis. In its idiopathic form, pulmonary arterial hypertension is a rare disease with an incidence of about 2–3 per million per year; other forms, however, are far more common.

Sildenafil relaxes the arterial wall, leading to decreased pulmonary arterial resistance and pressure. This, in turn, reduces the workload of the right ventricle of the heart and improves symptoms of right-sided heart failure. Pfizer conducted three trials on sildenafil in pulmonary arterial hypertension; the largest was the so-called Super-1 trial, an international multicentre, randomised, blinded, controlled study involving 278 patients with the disease. Based on the results showing improvements in exercise capacity, the company submitted an additional registration for sildenafil to the FDA, and it was approved for this indication in 2005. The dose of sildenafil required to treat pulmonary hypertension was as low as one-fifth of the dose for erectile dysfunction. The smaller, white pill was brand named Revatio™, and differentiated from Viagra by colour, shape, dose of active ingredient and by cost. As well as containing less active ingredient, it was a lot more expensive, because the disease is more serious and much rarer, so the costs of conducting the clinical trials needed to be shared among fewer prescriptions.

Like finasteride, this is an example of a large pharmaceutical company developing two products from one active ingredient for two different uses. Like finasteride too, the difference in the products is based on dose. A final point about sildenafil is that the various actual and proposed uses of the drug all involve increasing the supply of blood to particular parts of the body, namely, the penis and the lung (where it worked) or the heart (where it didn't work). Despite being in related fields, it is often unpredictable whether the medical hypotheses that underpin these types of developments for multiple therapeutic uses actually work in the real world.

Doxycycline: from killing bugs to protecting gums

The tetracycline antibiotics were first discovered in the 1940s, although doxycycline itself – one of the most successful of the group – did not arrive until 1967. The tetracyclines are broad-spectrum antibiotics derived from soil bacteria *Streptomyces aureofaciens*, so we have the biodiversity of the natural world to thank for this product. Doxycycline is a semi-synthetic analogue of one of the natural products. They have activity against many microbes including gram-positive and gram-negative bacteria, chlamydiae, mycoplasmas, rickettsiae and protozoan parasites. The mechanistic basis for their antibacterial effect is that they inhibit protein synthesis by preventing the attachment of aminoacyl-tRNA to the ribosomal acceptor site.

However, it took over 20 years from the discovery of doxycycline as an antibiotic for researchers at the State University of New York to find that it was able to inhibit the breakdown of gum tissue in periodontitis [57]. Gum disease is responsible for loss of more teeth than decay, and it is generally caused by an inflammatory reaction against the bacteria in dental plaque. Periodontitis is an inflammation of the periodontium, that is, the tissues that support the teeth. During the inflammatory reaction against the bacteria, the body releases certain enzymes (matrix metalloproteinases) which break down the gum tissue. The progression of the disease therefore depends partly on the host response, rather than solely on the cleanliness of the teeth. What was surprising was that doxycycline could inhibit matrix metalloproteinases, such as collagenase, at sub-antibacterial doses. This meant that low doses of doxycycline could be applied for the chronic treatment of periodontitis without stimulating the growth of resistant bacteria.

The academic research was taken up by the biotech company CollaGenex, based in Wayne, Pennsylvania, who developed a tablet called Periostat™ containing a sub-clinical dose of the antibiotic doxycycline hyclate for the treatment of periodontitis. In a Phase III clinical trial of 190 patients treated with Periostat over 9 months, there was a significant improvement in periodontal disease parameters. The product was first approved in the United States in 2001.

This is an interesting and unusual secondary use of an existing drug insofar as it involves a new mechanism of action. All of the other examples in this chapter concern a new application of a product but using the same original biochemical mechanism. In this case, the original effect of doxycycline against aminoacyl-RNA has been replaced by a more potent effect against matrix metalloproteinase enzymes. It was also unusual because it involved an R&D programme that was undertaken solely by a biotechnology sector company, with the product being taken all the way through to the market by that small company. One of the reasons CollaGenex was able to do this was because it was not starting with a new drug, and the overall development programme cost only $6.2 million. This is a very small fraction of what large pharmaceutical companies typically spend on innovative new products.

Raloxifene: from cancer to bone disease and back again

Raloxifene is a selective oestrogen receptor modulator (SERM) from Lilly, which was launched in the United Kingdom under the trade name Evista™ for the prevention of osteoporosis (brittle bone disease) in postmenopausal women in 1998. However, it had a surprisingly tortuous path to this endpoint, and even more surprising sequel to its primary purpose in terms of secondary uses.

SERMs mimic oestrogen in some tissues and have anti-oestrogenic activity in others; they are based on tamoxifen, which was discovered in the 1950s and used from 1973 for the treatment of breast cancer. The problem was that some patients were either resistant to tamoxifen, or became so some time after starting treatment, and Lilly was interested in finding drugs which would prove effective in combatting breast cancer in tamoxifen-resistant patients. They developed an analogue called raloxifene and tested it in a small 14-patient study, looking specifically for benefit in tamoxifen-resistant breast cancer patients. Unfortunately, it showed a poor response, and they decided to abandon their pursuit of this indication [58].

Rather than give up entirely, however, they were attracted to the development of raloxifene for osteoporosis. This condition is responsible for more than 8.9 million bone fractures annually among 75 million individuals in Europe, United States, and Japan with the disease, approximately 80% of whom are women [59]. Evista (raloxifene) is a modulator of the oestrogen receptor, the importance of which hormone is suggested by the high prevalence of osteoporosis in postmenopausal women (when oestrogen levels decline). Currently, up to half of women over age 50 will break a bone due to osteoporosis in their lifetime.

Towards this end, Lilly conducted a randomised, double-blind, placebo-controlled trial of nearly 8000 women in which increases in bone mineral density and reductions in vertebral fracture risk were assessed. The analysis after 2 years of treatment showed that women taking raloxifene were 52% less likely to have a first spinal fracture. They used interim results from this study for the regulatory filing, which was granted a priority review by the FDA and subsequently approved for this indication in the United States in 1999.

Slightly later, Lilly also evaluated it for the *prevention* (as opposed to treatment) of breast cancer. Looking at a continuation of the osteoporosis trial, they determined the risk of newly diagnosed breast cancer in postmenopausal women receiving raloxifene for the treatment of osteoporosis. It appeared there was a reduction of about 60% in the incidence of newly diagnosed breast cancer in this group of patients.

By 2006, it was announced that raloxifene was as effective as tamoxifen in reducing the incidence of breast cancer in postmenopausal women at increased risk. Unlike tamoxifen, raloxifene doesn't exert oestrogen-like effects on the uterus, a difference that was built into the discovery programme; as a result, the incidence of uterine cancer in patients treated with tamoxifen is thought to be higher than those treated with raloxifene (it is also superior in terms of risk of cataracts).[1] Subsequently, in 2007, the FDA announced approval of raloxifene for reducing the risk of invasive breast cancer in postmenopausal women.

So, finally, raloxifene had come full circle. While initially a failure in the *treatment* of tamoxifen-resistant breast cancer, raloxifene had proven to be effective in both osteoporosis and secondarily in the *prevention* of breast cancer; in the latter, it had proven to be a superior treatment to tamoxifen, in terms of risk of side effects.

Galantamine: using snowdrops to improve memory

Not all pharmaceuticals come from a chemist's bench. Many are directly or indirectly the products of plants. The story of galantamine is that of a pharmaceutical with ethnobotanical origins, in other words deriving from folk medicine. It is a rare example of a drug that originated to the east of the Iron Curtain, from countries within the Soviet Bloc during the time of the Cold War.

In the mountainous Caucasus region of Southern Russia, the wild-growing common snowdrop was used to ease nerve pain, where local people were used to rubbing their foreheads with the leaves of the plant to relieve headache. The Latin name for this plant is *Galanthus caucasicus*. In the 1950s, Russian academic scientists pursued this ethnobotanic observation, isolating and structurally determining the plant's active constituent, which was given the name galantamine. This plant alkaloid was found to act biochemically as a competitive reversible inhibitor of the enzyme acetylcholine esterase, which is responsible for inactivating the neurotransmitter acetylcholine. Acetylcholine is the body's most widespread neurotransmitter, responsible for the sensation of pain, the enablement of movement and many of the cognitive activities of the brain [60].

The industrialisation of this research was undertaken in Bulgaria, where the local pharmaceutical company Sopharma developed and commercialised a technology to

---

[1]In a 4-year follow-up of a clinical trial comparing the two agents, there was similar efficacy in terms of the prevention of breast cancer, but there were 36% fewer uterine cancers and 29% fewer blood clots in women taking raloxifene than in women taking tamoxifen, although the difference is not statistically significant [48].

produce galantamine on a large scale. Initially, galantamine was used as a neurological agent in the treatment of conditions such as post-polio paralysis and myasthenia gravis. It was also approved and used in neuritis and neuralgia, in anaesthesiology to increase muscle relaxation and, thereafter, in other areas of medicine.

In the 1980s, research on Alzheimer's disease began to focus on inhibitors of acetylcholine esterase, which were found in small-scale clinical trials to improve memory in this condition. However, one of the problems with this approach was that the enzyme is so widely distributed around the body that indiscriminate inhibition causes severe side effects. In fact, irreversible acetylcholine esterase inhibition is the basis for the action of chemical weapons such as Sarin gas, which causes death in major part by paralysis of the respiratory muscles. Galantamine has two advantages with respect to safety over these agents: first, it is reversible, rather than irreversible; and second, it readily crosses the blood brain barrier and concentrates in the central nervous system. Indeed, confidence in the safety of galantamine was to a certain extent assured by the previous medical use in other conditions. Preclinical investigations of galantamine for use in Alzheimer's disease ensued, and clinical development for cognitive impairment began in 1985. Full-scale clinical trials were conducted in the 1990s, and galantamine was first used (as Nivalin™) for the treatment of Alzheimer's in 1996. Subsequent commercialisation of galantamine was undertaken by Janssen Pharmaceutica (a subsidiary of Johnson & Johnson) and Shire Pharmaceuticals Group. In 2000, the compound emerged under the new trade name Reminyl™, and in 2001, the FDA approved its use for the treatment of mild-to-moderate cases of Alzheimer's disease in the United States.

Cyclosporine: preventing immune attack on organs and skin

The discovery of cyclosporine as an immunosuppressant also has at its origin a natural product. In this case, the Swiss company Sandoz (now part of Novartis), based in Basel, had instigated a screening programme to identify immunosuppressive effects of various fungal and bacterial extracts. The main aim of the programme was to discover immunosuppressants which were sufficiently devoid of cellular toxicity to be developed. One of the main achievements of the team was to set up a single animal model to do so, thereby saving money and animals. In late 1971, a fungal extract containing cyclosporine as its major constituent was tested and found to produce considerable immunosuppressive activity and little adverse cellular toxicity [61].

The early discovery stumbled almost at the first hurdle, with an inability to repeat the screening results. However, the discrepancy was resolved after further purification of the fungal extract, and isolation and synthesis of the active principle. Cyclosporine was biologically characterised for its ability to prevent rejection of skin and bone transplants in experimental models in 1976. Shortly thereafter, the discovery was moved into the early clinical phase with collaborators based at Addenbrookes Hospital in Cambridge, and at the Royal Marsden Hospital in London, England. Cyclosporine was initially established to benefit patients with kidney and bone marrow transplants.

Sandoz took up the later studies required for regulatory approval, which was obtained in 1983, for the prevention of organ rejection in kidney, liver and heart transplants. The discovery was of major medical importance, as it provided much greater success rates for organ transplants, which until then were poorly tolerated as a result of rejection by the host body as foreign tissue. Beyond this indication, however, the company also realised that an overactive immune system underlies many autoimmune diseases, including rheumatoid arthritis and severe psoriasis.

Subsequent to the approval for transplant rejection, Sandoz developed cyclosporine for both these indications, obtaining FDA approvals in 1997. A further development was undertaken for a topical ophthalmic formulation of the drug for another autoimmune disease, keratoconjunctivitis sicca, and this was approved in 1999. Although little used for rheumatoid arthritis, cyclosporine provided some impressive results in severe psoriasis, sometimes resulting in total clearing of the condition. Cyclosporine is a powerful drug however, and safety issues remained, since it increases the risk of infection and other health problems, including cancer; subsequent innovations in the past decade took the standard of care in this disease to newer, preferred therapeutics. So, yet again, in cyclosporine, we have seen the utility of a drug originally proposed for one use being actually useful in a variety of uses. Many of these are related, but as we saw with sildenafil, not all of the proposed uses turn out in practice to be workable. Finding out which hypotheses are valid and which are not requires experimental trials.

Dimethyl fumarate: a remarkable drug for multiple sclerosis

In 2013, a compound we had known for 200 years was approved both in Europe and in the United States for the treatment of relapsing-remitting multiple sclerosis (MS). The product manufacturer was the large biotech company Biogen Idec, who had licensed it from a small German company called Fumapharm.

The product, code named BG-12, is more commonly called dimethyl fumarate, known since the early days of organic chemistry and first synthesised as early as 1819, so it is nearly 200 years later that the use in MS was approved. How is it that today's pharmaceutical industry, with its high-throughput screening and computer modelling, cannot have identified this important use before now? My own view is that this is due to the fact that dimethyl fumarate was seen for at least 150 of those 200 years only as an organic chemical rather than as a pharmaceutical, without plausible biological effects. But if my view is correct, then there are likely to be many other uses for existing compounds which we do not yet know about.

Dimethyl fumarate was used for a long while as a mould inhibitor, for incorporation in leather items such as sofas during storage. However, it is also an allergic sensitiser at very low concentrations (down to one part per million), producing extensive, pronounced eczema that is difficult to treat. In 2007, 60 Finnish users of leather sofas that had been manufactured with dimethyl fumarate in them had serious rashes, and as a consequence, the importation of products containing dimethyl fumarate has been banned in the European Union since 2009.

Prior to consideration for the treatment for MS, dimethyl fumarate (and other fumarate esters) had been used for the treatment of psoriasis for decades, and a product called Fumaderm™ had been approved in Germany for this use since 1994. Both psoriasis and MS share immune and inflammatory mechanisms in their pathophysiology. This is not the same as saying that the use in MS is obvious based on the prior, known use in psoriasis, but nevertheless, the two indications are related, and some therapies (e.g. steroids) are used in both.

Dimethyl fumarate's mechanism of action involves up-regulation of nuclear factor (erythroid-derived 2)-like 2 (Nrf2) and subsequent induction of an antioxidant response. It does this by modifying the cysteine groups of a protein called KEAP-1, which tethers Nrf2 in the cytoplasm. Once modified by fumarate, the KEAP-1/Nrf2 complex dissociates, and Nrf2 migrates to the nucleus, where it activates various antioxidant pathways.

While Fumaderm was a popular product for psoriasis within Germany, Fumapharm did not attempt to gain approval in other countries. However, Fumaderm was found to have a general effect on the immune system, since in addition to an effect in pso-riasis, it also had a benefit in various other autoimmune diseases, advantageously so via an immunomodulatory effect rather than an immunosuppressive one. As a small company, Fumapharm looked for a larger partner to develop the wider commercial opportunity of their product. Once licensed to Biogen Idec, this new company was able to devote much more firepower to the development, including the much longer-term studies necessary for MS. This became their preferred therapeutic focus, and they branded the product Tecfidera™. They sponsored two main trials to prove the efficacy of the product involving 1200 and 1430 patients with relapsing-remitting MS, conducted over 2 years.

Tecfidera is an oral product, containing a slightly different composition of fuma-rate esters at a higher dose than Fumaderm. Oral products have a great advantage over the injectable MS alternatives like beta-interferon, but it arrived on the market around the same time as two other oral products for MS. Tecfidera has advantages over these too, with less risk of cardiac adverse events compared with fingolimod, and less risk of liver toxicity compared to teriflunomide. Although dimethyl fumarate is an old, well-known compound, it seems that its commercial outlook as a pharma-ceutical is bright. Now approved in both the United States and Europe and having recently been allocated 10 years of regulatory exclusivity in the latter territory, the forecasts are for blockbuster revenues [62]. This is a remarkable story of the intro-duction of a valuable new therapeutic option in a very serious disease that has effectively lain right under our noses for many years.

Botox: a drug to kill or cure you

Botulinum toxin, better known by its trade name Botox, is a bacterial toxin first dis-covered in poorly prepared sausages during the eighteenth century – the Latin for sausage is botulus. It is produced by the anaerobic bacterium *Clostridium botulinum*. Botulinum toxin is highly toxic; given intravenously, the lethal dose is of the order of 1 millionth of a gram per kilogram. With this background, it is remarkable that botulinum toxin has any useful pharmaceutical uses at all; in fact, it has a wider set of uses than any other approved drug.

Botulinum toxin kills its victims by causing respiratory paralysis. It is a neuro-toxin, and its mechanism involves the inhibition of protein formation in cholinergic neurons and thereby inhibition of the release of the neurotransmitter acetylcholine at the neuromuscular junction. Restoration of muscle function requires the growth of new nerve endings, making the biological effect of botulinum toxin very long lasting.

In the early 1970s, botulinum toxin was shown to relax musculature: a San Francisco ophthalmologist, Alan Scott, used it for the treatment of crossed eyes (stra-bismus) and uncontrollable blinking (blepharospasm). Academic investigators in the United States and Canada did further work to refine the injection protocol in the 1980s. Allergan, then a small biotechnology manufacturing company, took this for-ward, developing a specific form of the toxin, onabotulinumtoxin A, for these indi-cations and obtained the first approval by the FDA in 1989 for their trade named product Botox.

Allergan subsequently systematically investigated it for a wide range of new therapeutic uses, from wrinkles and hyperhidrosis (a chronic condition of excessive sweating), through to cervical dystonia (a neuromuscular disorder involving the head and neck) and spasticity. Botox is now widely approved by regulatory agencies around the world. In the United States, there are a total of 10 licensed indications; there are multiple other indications which are still experimental, off-label uses. The

FDA approvals are for chronic migraine, upper limb spasticity, cervical dystonia, hyperhidrosis, blepharospasm and strabismus, two types of facial wrinkles and two bladder dysfunction indications. New uses continue to be reported and investigated. Allergan's sales of Botox were expected to reach $2 billion in 2013. It is a remarkable example of versatility in a pharmaceutical product that was first known as a lethal poison in rotten meat.

With Botox, we have started to discuss drugs which have been approved for one or more indications but have off-label or investigational possibilities for many more. We can now develop this further in the next section.

Examples where evidence is uncertain and not to regulatory standards

Tricyclic antidepressants: for curing more than emotional pain

Quite a few of the earlier examples delineate a determinate discovery process that ultimately delivered two products based on the same agent, for separate uses. One of these products was used at a higher dose than the other, or in a different formulation, perhaps even by a different route of administration, but both were fully developed and approved for their respective uses based on widespread clinical trials.

In the majority of other cases, it is one product rather than two that has covered both primary and secondary therapeutic uses. Serendipity has often guided the identification of wider uses of the product after it has been initially marketed. Pain is one area where this is relatively common, the reason being that it is found either directly as a result of the condition being treated or as an associated condition, and something that a patient is likely to report to a doctor. For instance, pain and depression often go together. Indeed, patients often do not self-diagnose themselves with depression and instead first visit their primary care physician with somatic complaints, such as vague, nonspecific pain (e.g. headaches, muscle aches), feelings of malaise, decreased energy or insomnia.

The increasing and evolving use of antidepressants over the last 40 years has also brought about significant changes in the treatment of other conditions that co-occur with depression. One of the older types of antidepressant were the tricyclics, which were introduced in the 1960s and are functionally different from the newer selective serotonin reuptake inhibitors (SSRIs) like fluoxetine (Prozac™). Observant clinicians quickly noticed that certain tricyclic antidepressants such as amitriptyline were effective for the relief of chronic pain conditions. This was followed up by clinical trials to confirm that amitriptyline is indeed useful in painful conditions such as diabetic neuropathy and post-herpetic neuropathy, whereas conventional analgesics such as opiates or non-steroidal anti-inflammatory drugs are poorly effective or not effective at all. The onset of pain relief normally occurs at a lower dose and at an earlier time point than the antidepressant effect. This favourable quality is important because it means the patient is not exposed to safety issues that would otherwise accompany a higher dose, or longer time of administration relative to the dose or chronicity already established for the treatment of depression.

Amitriptyline and other tricyclic antidepressants are supported for their efficacy in the treatment of neuropathic pain by a wealth of clinical trial evidence [63]. Information gathered over several decades suggests that tricyclic antidepressants are effective for the treatment of persistent neuropathic pain. This support, based on good quality, patient-oriented evidence from randomised controlled trials, is focussed on diabetic neuropathy and post-herpetic neuropathy. There is also evidence in other forms of neuropathic pain such as from stroke, MS, migraine or tension headache. However, amitriptyline is not effective for the relief of acute arthritic pain. It has also proven unsuccessful at treating neuropathic pain from spinal cord injury. So, again, the exact clinical utility is not always predictable and requires experimental trials to find out exactly when it is effective and when it is not.

The pain-relieving effect of antidepressants is thought to be due to a combination of their pharmacological effect at inhibiting the uptake of both serotonin (5-HT) and noradrenaline into nerves, thereby making more of these neurotransmitters available in the neuronal gap (synapse) for their post-synaptic effect. Comparisons of the effect of drugs that act against both of these uptake processes suggest that the dual-acting agents, like amitriptyline, are more efficacious than ones which act solely on one system, like the SSRIs (e.g. fluoxetine (Prozac)).

Despite all this evidence, tricyclic antidepressants are not generally approved for these uses by regulatory sanction. This is mainly because these products are generic and it does not make commercial sense for the manufacturers to invest in proving safety and efficacy for their analgesic effect to regulatory standards. It is a shame since regulators give us gold standard approval status, and we expect our pharmaceuticals to have undergone their scrutiny. Medical professionals therefore rely on the ability to prescribe off-label for a particularly well-supported set of uses of these drugs. If the same wealth of evidence were true of all other off-label prescriptions, there would be little point in writing this book, but it is not true.

In addition to the antidepressants, there are also some anti-epileptics with a range of secondary and tertiary uses; some of these are approved but others are not. Topiramate (Topamax™), for example, was developed as an anti-epileptic drug by Ortho-McNeil (part of Johnson & Johnson). In addition to being approved both as monotherapy and as an adjunctive ('add-on') treatment for this indication, it is also approved for the prophylactic (i.e. preventative) treatment of migraine. So far, so good. However, there are other off-label uses of topiramate including the treatment of essential tremor, bulimia nervosa, obsessive–compulsive disorder, alcoholism, smoking cessation, idiopathic intracranial hypertension, cluster headache, treatment-resistant depression, cocaine dependence, diabetic neuropathic pain and borderline personality disorder. This type of list, by its length, renders each individual entry suspicious. Even for the most well supported of these uses, alcoholism, a Cochrane review from 2008 stated that 'data are insufficient to support using topiramate in conjunction with brief weekly compliance counselling as a first-line agent for alcohol dependence' [64].

Then, there is bupropion. Originally, this was developed for the treatment of depression, with the first patent on this product being filed in 1969 by the UK-based pharmaceutical company Burroughs-Wellcome. Later, the company became Glaxo Wellcome (and, later still, GSK), and bupropion was reincarnated as a smoking cessation product. In the United Kingdom, the depression indication disappeared. In terms of its ability to help smokers quit, it is similar in efficacy to nicotine patches and holds regulatory approval for this indication. Later still, it was approved for seasonal affective disorder, a type of depression that has a seasonal pattern. It has also been used off-label for various investigational uses such as bipolar disorder, cocaine dependence, irritable bowel syndrome, obesity, attention deficit hyperactivity disorder and so on, without demonstrating convincing levels of efficacy.

As we shall see later, the evidence base behind many unapproved uses of prescription drugs is very thin, and the problem we need to address is how to prevent patient harm from poorly supported unapproved medicine while maintaining prescriber freedom to use well-evidenced unapproved medicine. We also need to maintain the ability for doctors to be able to treat patients in their best interests when there are no approved treatments, such as the leper who arrived in Jerusalem near to death and was successfully treated with thalidomide. Discoveries such as these enlarge the therapeutic options from our existing pharmacopoeia, and I believe there are many more to be developed; however, we don't want our medicine cabinets polluted with ineffective or unsafe or poorly documented therapies.

Aspirin for cancer

Aspirin is another excellent example of a drug with multiple uses; indeed, it is truly charmed. One of the oldest known therapeutics, it was first used as a painkiller and then became established for cardiovascular disease (such as stroke and myocardial infarction). Then again, in December 2010, we saw the release of important clinical information supporting the benefit of aspirin in preventing colorectal and other cancers [65]. One could hardly have scripted a group of clinical benefits of greater medical importance.

Aspirin derives from the bark of the willow tree, the use of which to relieve headaches, pains and fevers was known to Hippocrates. Its isolated form was discovered in the early nineteenth century, and Bayer aspirin first appeared on the market in 1899. Much later, in the 1960s, the British pharmacologist and Nobel Laureate John Vane discovered in his laboratory at the Royal College of Surgeons in Lincolns Inn, London, that aspirin could disrupt a pathway needed for platelet aggregation. Further studies in the 1980s showed that this effect could be used for the prevention of heart attacks and stroke. Nowadays, low-dose aspirin is used widely for this effect. Beyond cardiovascular properties, there was suspicion that the same pathway could be involved in cancer. A further 30 years was required for the amalgamation of this clinical evidence in colorectal cancer; we are now nearly 200 years after the first use of aspirin as a painkiller, showing how long it can take for therapeutic substantiation even with a well-known drug. Indeed, even the paper from 2010, based on data from 25 000 patients, is published with the caveat that 'further research is needed'.

One of the problems with this discovery is that this research involves the use of aspirin to prevent cancer in a patient who might not get cancer at all. In people at low risk of colorectal cancer, the notable gastrointestinal side effects of aspirin are of concern. It is difficult for the medical professional to know whether to prescribe (or for the individualised patient to self-prescribe) the drug without a detailed knowledge of the risks and benefits. The use of aspirin for colorectal cancer has not been assessed by regulators, so it is still an 'off-label' proposition, as are the remainder of the examples in this chapter.

Retrospective data: looking back to create future therapies

One promising way of finding new uses for drugs is to analyse outcomes from patients. I have described how pain relief can be identified among antidepressants because of spontaneous reporting and observant clinical analysis. A much deeper type of analysis can be deployed to look for the effects of drugs on patients outside their intended therapeutic value. We are fortunate in the United Kingdom in having one of the world's biggest collections of patient data, called the Clinical Practice Research Datalink (CPRD), based on the experiences of the NHS in the 60 years since it was established. The CPRD is a rich source for retrospective analysis of this kind [66]. Cancer protection is an area which is particularly likely to be identified as an unintended benefit of a treatment for the reason that cancer incidence is a possible indication of carcinogenicity of a drug and should therefore always be recorded. If the incidence of cancer for patients on a particular drug is less than would be anticipated by chance, the drug may be exerting a protective anti-cancer effect.

Retrospective analysis, from what are also called 'observational' studies, is a powerful way of finding associations in real patients. But although this type of analysis, which I will refer to later in the book, is extremely valuable at suggesting relationships, it does not prove them and in particular does not establish their causal nature. For that, you need a prospective study in which the investigator declares in advance what is being tested and then conducts the experiment. The difference is rather as in the game of pool, when the player has to declare which ball she/he will pocket and which pocket it will fall through. The declaration in advance clearly reduces the chance of a lucky strike, in which, say, the primary target ball bounces off three cushions and hits a secondary ball into a pocket at the opposite end of the table from that intended. In pool, the harder the ball is hit, the more likely a ball will eventually fall through a pocket; so, in observational analyses, the more measurements that are taken, the greater the probability that a false positive relationship will show up.

With this caveat, a number of interesting associations have been found, and drugs as diverse as metformin (an anti-diabetic), propranolol (an antihypertensive) and clomipramine (an antidepressant) may have useful anti-cancer properties. Metformin in particular has raised hopes not just for preventing, but also for treating cancer, and a number of trials have begun for the use of this drug, including a large-scale adjuvant study in breast cancer [67].

These are but a random selection of interesting examples where secondary uses have been found. The list is not endless, but I will exhaust the reader's patience by going further. It is much easier to propose the secondary utility of an agent than to demonstrate it convincingly; in the middle between convincingly proving a secondary use and convincingly disproving it, one may demonstrate utility but unconvincingly. And some surprising failures can be discovered for combinations of drugs and uses that are very close to successful pairings.

CHAPTER 3

Shared decision making and consent

Off-label use becomes an ethical issue when the principle of informed consent is introduced. Those ethics are, to a certain extent, built into the professional codes of conduct by which doctors are bound. In some places, they are also embodied in legislation, such as in the European Union. This legal and ethical position derives from the history of informed consent, which arose in response to the medical research abuses embodied in the mid-1930s in Nazi Germany, through to the mid-1970s in mental health wards in the United States. Simply put, informed consent demands that patients give their consent to any treatment or research protocol that a clinician proposes.

The first point to make about off-label medicine in this regard relates to the massive public ignorance of the practice and some distrust. According to a poll of just over 3000 US adults conducted for the *Wall Street Journal* in 2006, around half wrongly believed that a doctor can prescribe drugs only for the diseases for which they have been approved. In addition, a 48–27% plurality believed that doctors should not be allowed to prescribe a drug for regulatorily unapproved uses. And 62% of respondents agreed with the proposition that prescription drug use for unapproved medical conditions should be prohibited except as part of the clinical research trial [68]. These 2006 results confirm the results of a similar survey conducted in 2004 [69]. Actually, it is not surprising that the public's views on the FDA regulatory status of the medicines they are prescribed do not tally with reality. It is not surprising because, in most cases, prescribers do not tell patients about the regulatory status of the prescriptions they write.

Patients are sceptical that a drug that has not been formally approved for a particular use, in a particular patient population, will indeed work in the way a prescriber intends. A study of 1000 members of the public indicated that the majority of adults believed that off-label prescribing to children would compromise safety and increase the likelihood of adverse effects [70]. A study by the same research group of children's views on the unlicensed use of medicines found they regarded the practice as unsafe and unethical [71]. Patients may be suspicious of pharmaceutical company over-claiming (Chapter 4), or the extent of the evidence in support of the new use (given that they know most drug trials fail), or generally concerned that the safety profile in the original indication is of concern in their particular case (Chapter 5). Public concern particularly applies when a patient's welfare is the responsibility of another person, when the patient is too young or too old or mentally incapable of making a rational, informed decision by themselves.

The 'informed' part of the term forces us to ask: how much information must patients receive in order to be able to give 'informed' consent? The answer to this question depends on who you ask; doctors and patients often have different views on the matter. In Lawrence Durrell's epic tome, *The Alexandria Quartet*, he recounts one

Off-label Prescribing – Justifying Unapproved Medicine, First Edition. David Cavalla.
© 2015 John Wiley & Sons, Ltd. Published 2015 by John Wiley & Sons, Ltd.

story from four viewpoints. It is a famous book because the true picture of a series of events that took place in the eponymous Mediterranean port only becomes clear to the reader after each of the participants' views of the events has been put forwards. As an area of controversy, off-label use is subject to the contradictory expectations of various stakeholders in the delivery of therapy, from producer to patient. Before addressing the issue of consent, we should briefly consider who these stakeholders are, since I will refer to them in future chapters also. Like *The Alexandria Quartet*, in this book, there are also four perspectives, all beginning with a 'P'. They are the patient, the prescriber, the producer and the payer.

The correspondence between the patient and the payer deserves some reflection. Unlike most commercial products, the person who receives the product does not normally pay for it, or at least not directly. In countries with socialised medicine, like the United Kingdom, the taxpayer funds the National Health Service (NHS), and of course, taxpayers are either patients or prospective patients, albeit ones that have not yet become ill. However, there is an imperfect correlation between patient and tax-payer, since in a general sense poverty is associated with ill health (and illness can reduce one's earning power) and in a statistical sense most prescription drug costs are related to the treatment of older people of pensionable age – in other words, they are net consumers of healthcare costs at a time in their life when they are not best able to pay.

Similarly, in the United States, Medicare beneficiaries, most of whom are aged 65 and over, made up 14% of the community population in 2001 but accounted for 41% of prescription medicine expenses. Over the period from 1992 to 2001, the number of prescribed drugs per patient per year increased by about 50% [72]. As we become an increasingly medicated society (whether for good or ill is a debate for another day), we can expect the number of prescriptions allocated to older people will also increase. So, while taxpayers as a whole pay for the drug bill, and government writes the cheques, the patients are heavily skewed towards people of pensionable age. In insurance-funded systems, the relationship between patient and payer is better correlated since premiums reflect likelihood of becoming ill and making a claim. But again, it is the insurance company that cuts the cheque. Whether correlated or not, the perspectives of patient and payer are not the same. The points of view of the patient, who almost always wants the best treatment, and the payer, who wants cost-effectiveness, are an important difference.

I now want to focus on the viewpoints of the patient and the prescriber, and how they see off-label medicine. In Chapter 4, I shall deal with the perspective of the producer; the payer is reserved for Chapter 2.

Viewpoint of the patient

With that background, let us now look at the viewpoint of the patient as regards consent. Most of the previous analyses of the issue of consent have approached it from a theoretical or philosophical perspective. It is actually difficult to understand consent in an abstract sense, unless we have been the victim of a situation where our consent has not been obtained, or even sought. To comprehend it better, we can go back to a scandal that was publicly revealed in the United Kingdom in 1999, at the Alder Hey Children's Hospital in Liverpool.

The scandal involved the unauthorised removal of human tissue from patients that had died at the hospital. In addition to the unauthorised removal, the organs were then stored, and often later disposed of without the authorisation of the patients' families. The issue was particularly inflammatory since in many cases the patients were children. The scandal was revealed through an inquiry into an initially unrelated specific case at the Bristol Royal Infirmary, spurred by accusations of excessively high rates of mortality from paediatric heart surgery at the hospital. The subject was the death of an 11-month-old baby girl, who had died while undergoing heart surgery; the inquiry occurred about 4 years after the baby's death and revealed that the girl's heart had been retained by the pathologist who performed the post-mortem.

One of the dead girl's parents set up a support group with other similarly affected parents, called the Bristol Heart Children Action Group, and then campaigned for a broader inquiry as to the extent of the practice throughout the United Kingdom. In 1999, the Action Group demanded that the public was made aware of the retained organs. During the subsequent public inquiry, conducted by Mr Michael Redfern, QC, it was revealed that various hospitals routinely kept organs of patients after their death; in particular, the Alder Hey Children's Hospital in Liverpool had kept a large number of children's hearts. One pathologist (Prof van Velzen) reported that he had *systematically* ordered the stripping of *every* organ from every child who had had a post-mortem during his time at the hospital, even in cases where the parents had specifically requested that there would not be a post-mortem. It emerged that, in addition to the Alder Hey, the Birmingham Children's Hospital had also sold thymus glands, removed from live children during heart surgery, to a pharmaceutical company for research. Alder Hey also stored without consent 1500 foetuses that were miscarried, stillborn or aborted. Walton Hospital, also in Liverpool, had stored the organs of 700 patients. When finally published as the Redfern report in 2001, the inquiry revealed a practice that operated over the period from 1988 to 1995, during which over 100 000 organs, body parts and entire bodies of foetuses and stillborn babies, in addition to nearly half a million tissue samples, were held.

The widespread nature of the practice aroused public outcry against the NHS, and the UK government was forced to respond to prevent a recurrence. In 2004, it did, with an overhaul of the handling of human tissues in the United Kingdom, the creation of the Human Tissue Authority and the legal passage of the Human Tissue Act; in the act, there was a specific new requirement for informed consent to be obtained for the removal, storage and use of human tissue [73]. There was also an attempt to prosecute Prof van Velzen, the pathologist at the heart of the scandal, but this failed in 2004 on a technical ground, in that the preserved tissues could not be assigned unequivocally to the post-mortem babies; this prosecutional failure then aroused further public consternation, and Prof van Velzen was permanently banned by the General Medical Council (GMC) from practising in the United Kingdom in 2005 [74].

To be clear, gruesome though the scandal was, Alder Hey involved a practice with solely ethical and no therapeutic concerns.[1] There was no accusation of a negative medical outcome due to the organ removal, since, after all the organs were only removed after the patient had died: the organ scandal was based on the offence

---

[1]Though there was an attempt to prosecute Prof van Velzen, in relation to the practice of disposing of dead bodies and body parts without consent, the offences that were considered were of misconduct in a public office and dishonesty. The possibility of a crime having been conducted under the Burial Act 1857 (which was used to prevent exhumation for the study of anatomy) was not considered since there had been no disinterment. Under English common law, a corpse is not recognised as a property belonging to any living person, making moot a prosecution based on theft.

caused to the surviving parents, a feeling of enormous disrespect shown to their deceased children. (Separately, there were concerns about excess mortality in some of the hospitals concerned, but that was a matter entirely unrelated to the organ-storing practice.) On the contrary, it could be argued that the retention of these organs was designed in various ways to further medical science.

In comparison, the public are currently mostly unaware of the practice of off-label medicine, but it is vastly more prevalent than the practice of organ retention. Patients who are prescribed off-label medicine without their knowledge have been misled. It was previously estimated that there are around 2 billion incidences where off-label prescriptions have been written in the seven major pharmaceutical markets each year. This is a massive issue of trust in healthcare provision.

Unless the patient is not sentient or able to express an opinion, or mentally inca-pable, it must be presumed that the patient would want to know (and indeed should know) about the off-label status of his or her prescription, since it would be perti-nent to the safety of their medicine. Apart from the unconscious emergency situation, these exceptions can largely be dealt with by seeking consent from the patient's parent, guardian or carer. In fact, in the United Kingdom, there is a specific requirement under the Mental Care Act 2005 for a health professional to consult with the carer of a person who is unable to consent. Without such consent, it is dif-ficult in the vast majority of situations to see how off-label medicine be prescribed in accordance with the Hippocratic Oath, which requires the doctor to 'keep his patient from injustice'.

Off-label medicine involves more than just consent for consent's sake: as previ-ously mentioned, there are also risks of poorer safety and efficacy outcomes. In consequence, the regulatory status of a prescription is a relevant medical fact. It therefore comes as a surprise to hear the medical profession object to greater trans-parency in the area of off-label medicine, with suggestions that a more regulated environment would lead to a 'tick-box' culture or otherwise 'box-in' medical prac-tice. To me, concealing the nature of an off-label prescription is diametrically opposite to a full and frank doctor–patient relationship and suggests an arrogant view which has not moved beyond 'doctor knows best'.

Not only is today's patient generally better informed and interested in their own healthcare and outcomes of treatment, they can easily acquire more information. Barriers between the medical professional and the layperson are broken down as oblique medical acronyms are decoded and obscure medical terms translated. The scale of medical information available to the patient is suddenly huge since the development of the Internet. Patients can also find out much more about their condition, its prognosis and treatment from Internet search engines. Moreover, they can do so by framing natural language search queries, requiring no special knowledge, and can do it from their handheld smartphone, as soon as they leave the doctor's surgery (or even in it!). They can also research their doctor and his or her track record. Google searches for healthcare search terms are ranked as some of the most frequent topics for research. European market research indicates that 75 million users accessed the Internet for pharmaceutical information in 2010, with Wikipedia being the most visited health resource.

As a consequence, doctors today are expected to engage with patients who are much more informed about their own health, care and treatment, in a reshaped rela-tionship [75,76]. Sometimes, true, the information patients have gleaned can feed a level of hypochondria which is unhelpful for the medical consultation (it's even got a name, 'cyberchondria'). Doctors may then need to instil a balance in the patient's knowledge, rather than to deny or ignore the existence of this inaccurate information.

But unhelpful situations aside, the Internet is not just for patients; doctors can also benefit, sometimes also undertaking their own healthcare information research!

Web-based organisations have sprung up to allow patients to talk to each other, without the intermediacy of the doctor, such as PatientsLikeMe (http://www.patients likeme.com/). This organisation was established to connect patients with the serious condition of amyotrophic lateral sclerosis (Lou Gehrig's disease), an area with only one approved treatment (riluzole) which is not very effective, and therefore high rates of off-label use of other agents. Users can view more than one option for their treatment and make their own choice based on the views and reports from 'patients like me'. As patients take charge of their own illness, the importance both of doctors and of pharmaceutical companies diminishes. Jamie Heywood, who founded the organisation, says that 10% of their users fired their doctor as a result of using PatientsLikeMe [77].

Today, a patient given a prescription of an antidepressant like amitriptyline for pain without being told it is for an off-label indication is likely to remark to their doctor with some irritation that they 'are not depressed'. This type of confusion can easily be avoided by the doctor explaining to the patient that although the antidepressant has not been formally approved for pain, there is a wealth of information supporting the benefit of the product in painful conditions [78]. Unfortunately, this wealth of evidence is generally not available for other off-label medicines – but this deficit of evidence makes it less, rather than more acceptable for a doctor to hide the off-label status from a patient.

Viewpoint of the prescriber

Set against the perspective of the patient is that of the prescriber. In the United Kingdom, the prescriber is most usually a general practitioner (GP); however, other medical professionals can prescribe drugs, including certain nurses, pharmacists, chiropodists/podiatrists, physiotherapists and radiographers.

The inclusion in this list of the pharmacist deserves some comment. The role of pharmacists was originally limited to dispensing prescriptions that were handed to them by doctors, but the present trend is for their responsibilities to be widened, so that they can also prescribe. In their traditional role, however, the prescriber is usually separated from the pharmacist. This practice began in the nineteenth century and confined the activities of the doctor to diagnosis and prescription. One of the perceived advantages was that it enabled any profit that could be made from the dispensing decision to be removed from the prescriber's motivation; another was that the prescription written by a doctor could be independently checked to avoid errors. This is the general position today in the United Kingdom and the United States, though there are some exceptions. It is different in other countries: for instance, in Canada, it is common for a medical clinic and a pharmacy to be located together and for the ownership in both enterprises to be common but licensed separately. In Asian countries such as China, Hong Kong, Malaysia, and Singapore, doctors are allowed to dispense drugs themselves.

Despite these complexities, we shall consider the ethical motivations of a prescriber as they would operate separately from that of a dispenser. So, taking one of the scenarios in Chapter 1, let's consider the prescriber's position in a case where the patient is a child. Paediatric medication poses particular challenges to the prescriber, if (as is often the case) there are no licensed drugs to treat the child's condition, only drugs that have been tested on adults. The professional bodies like the UK GMC have considered this type of situation and have issued guidance for doctors to follow in order to allow the prescription of an off-label medicine. The guidance makes it clear that the professional bodies do not regard this right to be unfettered.

Professional guidelines

The current GMC guidelines, released in 2013, were the subject of some discussion in relation to off-label use. The previous 2008 Guidance had limited the prescribing of off-label or unlicensed medicines to circumstances where there was no appropriately licensed alternative or where the doctor was satisfied that such a medicine would 'better serve the patient's needs'. The conditions were framed around situations that would frequently arise in relation to prescribing for children, rather than situations like the prescription of antipsychotics to Alzheimer patients or antibiotics for respiratory viruses. In 2011, the GMC was considering a new version of the guidance and proposed to relax the off-label guidance to allow off-label prescription on economic grounds to save money. They then put the proposed guidelines out for discussion. Some of the comments that came back, for instance, from the Royal Pharmaceutical Society, were critical of the changes to the rules regarding off-label prescribing. But more important, during the course of that debate, it turned out that there was an issue with regard to European law (as discussed further in Chapter 6), and as a result, the GMC reverted to the 2008 principles.

The resultant 2013 GMC Guidelines as they relate to off-label or unlicensed medicine come in three parts [79]. The first is a non-exhaustive list of the circumstances in which it is permissible to prescribe in this way; the second is the scientific basis for doing so, the liability that is taken on and the records that must be kept; and the third is the information that must be communicated to the patient in such circumstances.

In paragraph 69, the possible circumstances include the prescription of unlicensed medicines where there is no suitably licensed medicine that will meet the patient's need, for example, where

i there is no licensed medicine applicable to the particular patient. For example, if the patient is a child and a medicine licensed only for adult patients would meet the needs of the child; or

ii a medicine licensed to treat a condition or symptom in children would nonetheless not meet the specific assessed needs of the particular child patient, but a medicine licensed for the same condition or symptom in adults would do so; or

iii the dosage specified for a licensed medicine would not meet the patient's need; or

iv the patient needs a medicine in a formulation that is not specified in an applicable licence.

This paragraph also allows for situations where a suitably licensed medicine that would meet the patient's need is not available, for example, where there is a temporary shortage in supply. The term 'unlicensed medicine' is defined to 'describe medicines that are used outside the terms of their UK licence or which have no licence for use in the UK', therefore encompassing the term 'off-label' as used in this book.

Once a doctor is minded to prescribe an off-label medicine, they are then required, in paragraph 70, to:

a be satisfied that there is sufficient evidence or experience of using the medicine to demonstrate its safety and efficacy

b take responsibility for prescribing the medicine and for overseeing the patient's care, monitoring, and any follow up treatment, or ensure that arrangements are made for another suitable doctor to do so

c make a clear, accurate and legible record of all medicines prescribed and, where you are not following common practice, your reasons for prescribing an unlicensed medicine.

I shall be dealing later on in much more detail with the issues around evidence for off-label medicine (see p. 91), liability for medical professionals (see p. 105) and how off-label medicine should be recorded (see p.162).

The guidance then goes on, in paragraphs 71–73, to talk about the responsibility of the doctor to inform the patient when prescribing an off-label or unlicensed medicine:

71. You must give patients (or their parents or carers) sufficient information about the medicines you propose to prescribe to allow them to make an informed decision.

72. Some medicines are routinely used outside the terms of their licence, for example in treating children. In emergencies or where there is no realistic alternative treatment and such information is likely to cause distress, it may not be practical or necessary to draw attention to the licence. In other cases, where prescribing unlicensed medicines is supported by authoritative clinical guidance, it may be sufficient to describe in general terms why the medicine is not licensed for the proposed use or patient population. You must always answer questions from patients (or their parents or carers) about medicines fully and honestly.

73. If you intend to prescribe unlicensed medicines where that is not routine or if there are suitably licensed alternatives available, you should explain this to the patient, and your reasons for doing so.

Doctors are not regulatory agencies, and it would be invidious to ask them to take on that responsibility. I would therefore argue that paragraph 71 makes it unlikely that off-label or unlicensed medicine can be prescribed without conveying the absence of regulatory approval of the medicine for the patient's condition. I particularly note that this paragraph starts with 'You must' rather than 'You should'. It is difficult to see how a patient can make an informed decision if a crucial piece of information such as whether the regulatory agencies have sanctioned it is lacking, unless the doctor wishes to stand in place of the regulatory agency; as I say, this is undesirable. Even if a doctor has any doubt about whether it is important or necessary for the patient in making an informed decision, it is generally not up to him or her to withhold it, convenient though that might be.

Paragraph 72 then suggests some (rare) circumstances where such information may be withheld if off-label or unlicensed medicine is inevitable *and* if the knowledge of that information is likely to cause distress. A second exemption is where, despite the lack of regulatory approval, there is authoritative clinical guidance. We shall see examples of this emerging in the United Kingdom through the National Institute for Health and Care Excellence guidance on off-label uses (see p. 117). However, we can presume single case reports or recommendations from clinical colleagues would not fall into that category.

Finally, paragraph 73 makes it clear that these exceptions described in the previous paragraph do not apply if the off-label or unlicensed use is not routine or if there are alternative licensed medicines for the patient's condition.

Taken together, this leaves precious little room for off-label medicines to be prescribed without informing their patients that the drug has not been approved for the purpose they intend. Yet, a majority of medical professionals do just this. Various NHS Trusts in the United Kingdom have now issued guidance in regard to the policy and procedure for the use of unlicensed medicines, which states:

It is also important that when obtaining consent for treatment the prescriber should, where possible, inform the patient of the medicine's licensed status, in terms they can understand, and that for an unlicensed medicine that its effects will be less understood than for those of a licensed medicine.

Beyond the United Kingdom, this ethical position around informed consent has now been legally enshrined as a European right that all EU member states are bound to protect. In Australia in November 2013, the Council of Australian Therapeutic Advisory Groups issued a report advocating a new national framework for off-label prescribing which recommends obtaining written patient consent and closely monitoring treatment outcomes. In June 2014, the report was welcomed by a group of leading doctors [80]. However, as we shall see, the level of public awareness of off-label medicine does not match the frequency with which it occurs, suggesting that these ethical and legal considerations are often being ignored.

In the United States, the guidance from the American Medical Association (AMA) is less precise, having little or no specific reference to off-label medicines. The AMA is strongly in favour of the autonomous clinical decision-making authority of a physician and that a physician may lawfully use an FDA-approved drug product or medical device for an unlabelled indication 'when such use is based upon sound scientific evidence and sound medical opinion' [81]. However, in its Code of Medical Ethics, the section on Informed Consent (Opinion 8.08) says [82]:

> The physician's obligation is to present the medical facts accurately to the patient or to the individual responsible for the patient's care and to make recommendations for management in accordance with good medical practice. The physician has an ethical obligation to help the patient make choices from among the therapeutic alternatives consistent with good medical practice. Informed consent is a basic policy in both ethics and law that physicians must honor, unless the patient is unconscious or otherwise incapable of consenting and harm from failure to treat is imminent.

There is a principle of 'shared decision making' (SDM) in the United States, which is gaining increasing prominence in healthcare policy [83]. At its core, SDM aims to enable patients to be individually self-determinate in their treatment and operates on the principle that, wherever possible, clinicians need to support patients to achieve this goal. SDM involves more than just information transfer from prescriber to patient, extending to honouring the patient's informed preferences. It recognises a physician's duty to promote patients' well-being by openly discussing the balance between anticipated benefits of a given intervention and its potential harms. It involves a presentation of information about reasonable options in terms patients can understand. Finally, it involves both the doctor and the patient arriving at a mutually acceptable decision based on their shared knowledge and values [84].

Whereas informed consent describes a unilateral decision made by a client who has been provided all relevant treatment information, SDM is a concurrent, fluid exchange of information between patient and physician in a mutual attempt to reach a consensus treatment decision. SDM has made some progress in the United States in becoming a component of a range of healthcare delivery system reform initiatives at the federal level, most prominently through the Affordable Care Act of 2009 (the so-called 'Obamacare' programme). In addition to the federal initiatives, many states have taken steps to implement SDM through a variety of policy strategies, such as legislation and incorporation into state standards and expectations. States that are in the vanguard in this respect are Maine, Minnesota, Oregon, Vermont and Washington. In Maine, SDM is embedded in the process of licensure of physicians [85]. The Informed Medical Decisions Foundation (http://www.informedmedicaldecisions.org/) lobbies on behalf of patients for further measures to enhance patient involvement in treatment and issued a report on the policies and legal framework in 2012.

In the United States, as we discuss later (see p. 126), medical practice is not bound by regulatory edict. However, patients do not know this legal nicety, so in my view, while this may present a legal loophole for litigation lawyers to climb through, it does not change the ethical situation. Given that patients would generally regard the regulatory status of their medicine as a material 'medical fact' in making a decision on their treatment, the AMA Code of Ethics underlies an ethical prerogative that the patient should be informed of the off-label nature of a prescription. In other words, informed consent is governed by the information the patient needs to know to make an informed decision, not the information the doctor is prepared to provide. We will take up this issue again, when we discuss the legal issues surrounding SDM and informed consent further in Chapter 6, since physician liability is a corollary of consent, either as an explicitly mandated requirement or as a factor in deciding the outcome of a case for the ensuing damages.

Patient awareness

Given that in the United Kingdom, professional obligations generally require prescribers to inform patients when making an off-label prescribing decision, we might expect that the issues around off-label medicine would be well known to the public. For instance, it has been estimated that 40–76% of children have had off-label medicine (see p. 9). Logically, their parents should have been informed in the vast majority of these cases, assuming that the prescribers had been operating according to their professional guidance.

In fact, patient awareness is very low. A survey of 1000 members of the public in Northern Ireland indicated that 86% of the general public surveyed lacked knowledge about the unlicensed medicine use in children [70]. This is not because of indifference: 81% of patients would be concerned about unlicensed prescriptions [86]. When informed about this practice, the majority believed that it would compromise safety and increase the likelihood of adverse effects. It turns out they are correct in this belief. A very large majority (92%) felt that parents or guardians should be told if their child was prescribed a medicine that had not been fully tested in children. Only 18.2% indicated that they would 'simply accept that the doctor knows best'.

These statistics indicate that the reason for low awareness among patients is likely to be because they have not been told and presumably are under the impression that prescriptions are fully controlled by regulatory authorities, which govern not only medicines but also the circumstances in which they are used.

Practitioner attitudes

The Moonstone, by Wilkie Collins, involves at its crux a story of medical deceit. In one scene, laudanum is administered to the hero Franklin Blake by a Dr Candy to settle a wager but without Blake knowing. As a consequence, Blake unconsciously steals the moonstone diamond, a number of people die, and unravelling the mystery is the subject of the book. The fact that Blake is the subject of a treacherous act is the climax of the story, and not revealed until near the end. After a long series of false trails, Blake finally understands what has happened after tracking down Dr Ezra Jennings, Dr Candy's assistant. Jennings confesses to Blake that 'Every doctor in large practice finds himself, every now and then, obliged to deceive his patients'.

The Moonstone was written in 1868 and reflects a historic view of the ethical parameters of a doctor–patient relationship. We can hope that better standards apply nowadays. However, our hope depends on doctors being able to meet patients' expectations regarding transparency and honesty about the medicines they prescribe. We can be encouraged by the ethical guidelines in the UK GMC guidance referred to earlier, so long as prescribers are aware of, and adhere to them.

However, surveys of off-label medicine among prescribing doctors have produced inconsistent results that are difficult to explain. In a questionnaire surveying nearly 350 Scottish primary care practices, 74% of GPs admitted to being familiar with the concept and 40% to knowingly prescribing paediatric off-label medicine. However, less than half were aware that off-label prescribing is common in general practice, indicating a significant lack of information concerning the extent and nature of off-label prescribing in primary care. Respondents cited age-related reasons (as in paediatric medicine) as the most common basis for an off-label prescription, whereas other surveys have revealed dose to be the salient factor [87]. The high levels of awareness from this study about the issue also contrasts with the survey of GPs referred to earlier which found that only 14% of GPs were very familiar with the GMC guidelines [86].

Another study, again from the United Kingdom, was conducted among 1500 community pharmacists; these then are the people who fill the prescriptions rather than write them. From around 500 completed questionnaires, the authors of the study identified 70% professing to know about and 40% to having dispensed off-label medicine – similar levels to the previous GP questionnaire [88]. There were particular issues with the prescription of paediatric medicines outside approved dose ranges, such as high doses of inhaled steroids, β2-agonists or paracetamol and low doses of antibiotics (it is important to recognise that in this instance at least, the consequence of lower than approved doses can be even more serious than doses that are too high), and concern that some prescriptions may have been written in error or at least without proper awareness of the approved doses. Most community pharmacists stated that they should inform the prescriber that a medicine was off-label; however, when given specific practical examples, less than half would actually appear to do so.

A similar outcome was observed in a survey of 300 Jordanian doctors. Here, 90% of respondents professed 'concerns' about the safety and efficacy of paediatric off-label prescriptions, yet there were 28% who knowingly prescribed in this way. Of this proportion, only 10% of the off-label prescribers can be 'unconcerned', and sheer mathematics would tell us that at least 18% of the 28% would knowingly prescribe in a way that caused them concern, yet the survey also revealed that the majority do not convey their concerns to their patients [89].

There is a revealing account of off-label usage in German academic gynaecological departments, in which although one of the employed physicians in one hospital referred to off-label use as a common practice for director, assistant medical director and interns, the director of the same hospital denied off-label use. It is clear from this example at least that some doctors do in fact prescribe off-label but do not wish to be identified [90].

One possible, perhaps even a likely explanation for these conflicting statistics, is a low level of compliance among members of the medical professions with the obligations and guidance of their professional associations, along with a reluctance to admit to actual off-label prescription behaviour. The medical profession has sometimes seen fit to justify the concealment from a patient of the approval status of a medicine. Forms of justification vary from 'the truth might distress the patient' to 'the regulatory status is irrelevant' to 'the patient doesn't need to know'. A particularly strident defence of the physicians' right not to obtain informed consent is provided by the medical litigation defence lawyer Jeff Beck [91].

Beck's viewpoint hinges on a central falsehood that physicians' off-label choices regarding medicine for their patients are the correct choices, despite not having undergone regulatory review. He argues that off-label medicine is good medical practice that necessarily bypasses the torturous process of regulatory approval to the direct benefit of patients, because obtaining the evidence is costly and time consuming, but is not likely to deliver a different, unexpected result. He fails to point out that obtaining the evidence is also very likely to disprove a medical hypothesis or uncover a problem with safety that is not revealed with the small numbers of patients in a physician's direct experience. Inconveniently for Beck, most theories about new uses for existing drugs turn out to be either unsafe or ineffective, as revealed by a recent *New England Journal of Medicine* study on 363 long-standing medical myths, only 38% of which survived evidence-based analysis (see p. 101). An even greater rate of failure, at least 90%, is observed if one considers the likelihood of successful passage of a new treatment through the stages of clinical development and into clinical practice.

There is a strange parallel between the arguments in defence of a physician's decision to prescribe off-label and that to defend pharmaceutical off-label marketing. It is strange because there is a general belief that off-label pharmaceutical marketing is broadly undesirable and wrong (indeed, as I will deal with in Chapter 7, regulatory agencies have been avid in seeking to prevent it happening), whereas off-label pre-scriptions are imperfect but have their good aspects. Yet both arguments in support are based on a similar conjecture, that being 'what if it is true?' The ostensible truth of an off-label marketing claim has been used in US legal cases, most recently in support of a medical rep from Orphan Medical in a suit against the FDA. If the marketing claim is true, it may be supported by the First Amendment right of free speech (see p. 139). If it is false, it may not. Similarly, for those in support of off-label medicine, the essential unimportance of regulatory approval is ensured if the evi-dence backs the conjecture: 'it is true, the hypothesis is valid, who cares what the FDA says, it is irrelevant for the patient and their informed consent is not needed'. However, such outcomes are the exception rather than the rule: most hypotheses are wrong, and when they are false, defending them against inspection, analysis and gathering of evidence is very wrong indeed.

But for my scepticism about off-label medicine to be valid, I wouldn't even need the huge clinical trial failure rate that statistics show to be the case; I would only need a small chance that a hypothesis turned out to be wrong to justify it being tested. With 2 billion prescriptions written off-label annually, the scale of the practice is far too extensive for hand-waving arguments about 'doctor knows best'. Patients are no longer willing to countenance hypotheses affecting this many people without proof. This, after all, is the reason for the regulatory system which governs pharmaceuticals in the first place – a system which, incidentally, has imposed increasingly high hur-dles of patient safety in medicines as time has progressed. As we shall see in Chapter 5, even fairly minor deviations from the medicine's label can result in unexpected fail-ures, and these failures can derive not just from inadequate efficacy, but also inade-quate safety. There is no need to reserve the demand for evidence to the extreme cases where a pill for insomnia is repurposed for leprosy. There are cases like Fen–Phen (fenfluramine/phentermine), where safety dangers were uncovered without a change in use but just from using the product longer than the label entitled it to be used; or Seroxat™ (paroxetine), where other safety issues arose from its use in chil-dren, rather than adults; or Avastin™ (bevacizumab), which is effective in colorectal cancer but not breast cancer [92].

So, to summarise, I take a paragraph from Ghinea et al. [85]:

> Clinicians need to be conscious of the fact that they often prescribe off-label and that all such decisions can have serious implications – particularly when evidence of efficacy and/or safety is lacking. Once this has been acknowledged, clinicians will be in a better position to reflect on their own prescribing practices and ask themselves: 'Am I privileging the values of evidence-based medicine, or am I privileging other factors, such as common professional practice, clinical impression, physiological rationale?' And 'To what degree

has my decision to prescribe this agent been influenced by drug company marketing?' Whatever the answer, it is then the clinician's responsibility to acknowledge the implications of his or her decision for the patient as well as for the healthcare system and (re)consider whether their off-label use of this medication is justifiable. The clinician also needs to consider whether and how the patient is informed about off-label use – a complex ethical and legal issue given that some off-label uses are supported by substantial bodies of evidence whereas others are closer to being 'experimental'.

I couldn't agree more. The future does not belong to Beck: his arguments derive from a world of paternalistic attitudes to medical practice, configured to help his physician and pharmaceutical clients. In today's good medical practice, the rights of the patient take precedence over the convenience of the practitioner. Informed consent is increasingly an indelible component of the modern relationship between doctors and their patients. It reflects a change away from a historical situation, where the doctor exerts all the decision-making power and the patient remains passive, to a more consensual, mutually respectful relationship, where the medical consultation is focussed around the main character of the patient, who has basic rights. These rights are conditioned by the fact that they are ill and their physiology about to be altered, hopefully in a good way. Informed consent is the means to ensure that the beliefs of the doctor and other practical considerations do not override the patient's right to self-determination and personal integrity.

Diagnosis

Before leaving a chapter about 'decision making', we should mention the process of diagnosis, since it is central to a consideration of off-label medicine. A medical consultation revolves around a diagnosis, which then becomes the basis for a prescription (where necessary). We can find out whether a prescription is on- or off-label if we know the diagnosis, the patient's age, sex, etc., and the medicine written on the prescription.

Diagnosis or diagnostics is the process of identifying a medical condition or disease by its signs and symptoms and from the results of various tests or procedures. The term 'diagnostic criteria' designates the combination of symptoms which characterise the disease or condition. In some case, the diagnosis is only hypothetical, a 'provisional formula designed for action'; further testing is necessary to confirm or clarify the situation before providing treatment.

Diagnosis is particularly problematic in psychology or in psychiatric medicine, where objective measurements such as blood pressure or glucose levels are replaced by much more subjective criteria. Off-label prescriptions are particularly prevalent in psychiatry, as we saw earlier. By going into this issue in some detail, I hope to show up some of the problems that can arise in the real world, where marginal differences can determine whether a particular prescription is on- or off-label.

In psychiatric medicine, the *Diagnostic and Statistical Manual of Mental Disorders* is a standard text, now in its fifth edition. Known primarily by its acronym, DSM-5 presents itself as a manual for clinicians (a definition that encompasses the often competing schools of psychology, psychiatry, psychoanalysis and so on). There has been a major furore around the publication of DSM-5, since it concentrates mostly on symptoms rather than the biochemical or neurochemical basis for mental illness, which some believed should underpin diagnosis [93].

Part of the reason for the establishment of the DSM was the differential diagnosis of schizophrenia on either side of the Atlantic: in New York, there were about twice as many schizophrenics diagnosed in the early 1970s as in London. The DSM standardisation was meant to eliminate this arbitrary geographic difference. Since the cause of mental illness is still unclear, DSM-5 concentrates on symptoms, or behaviours, on which there could, in an ideal world, be universal agreement. As far as schizophrenia is concerned, in an earlier version of the DSM, 'serious psychosis' was subdivided into what is now known as schizophrenia and bipolar disorder.

In DSM-5, there are now numerous subtypes of schizophrenia, from bipolar I and II (the latter a milder form of the former) through to 'schizophrenia spectrum' and 'other psychotic disorders'. Other forms of mental illness are similarly subdivided by menu. For instance, acute stress disorder has two primary divisions, into A and B type. Each is defined according to the experience the patient suffered in becoming stressed and then further subdivided according to a list of symptoms. In addition to the complexity, a major difficulty lies in the fact that at the end of the menus, we are often left with a catch-all of 'not otherwise specified'. A further complication arises from the fact that many mental disorders co-occur; so 'catatonia' is often co-morbid with schizophrenia or anxiety disorder with ADHD.

The research-driven world complains about the arbitrary nature of these categorisations. Prof Tom Insel of the US National Institute for Mental Health (NIMH) seeks to define mental illness according to genes, molecular markers and so on. But we are a long way from this biochemical understanding of mental health, and until we have such an understanding, symptomatology is the best we have. The present methods of diagnosis, particularly in mental health, can require fine judgments to be made. The practitioner's job is difficult under these situations, yet the categorisations are necessary for the regulatory system which approves drugs on the basis of very specific sets of disease, and for the reimbursement of these medicines. I can see the difficulty the prescribing physician faces, and have some sympathy for his/her predicament, but all the other players in our medical system (pharmaceutical company, payer, patient and so on), have to fit in with these descriptions.

Diagnoses are not the result of an over-ebullient attempt to categorise patients for the sake of it. Diagnoses are the language by which clinicians can describe the patient to other clinicians, incorporate him or her into a clinical trial, and the basis on which a new medicine becomes approved by a regulatory agency. This then fashions the ontology by which real world patients are treated. Then, diagnoses are critical for categorising medicines as on- or off-label, from which a prescriber's legal liability can arise in case things go wrong.

Sometimes, patients are misdiagnosed, but errors in this regard are a fact of life, which may be reduced but cannot be eliminated. The purpose of this book is not to counsel for a perfect diagnosis in all cases. No, my purpose is quite different; it is to ask what are the consequences of a prescription that is not approved for whatever the doctor *thinks* is the correct diagnosis. Misdiagnosis is not the same as off-label medicine, which is the decision of what to prescribe after a diagnosis has been reached, be that diagnosis correct or incorrect. At the same time, it is undoubtedly true that if diagnoses are inaccurate, patients risk being treating with the wrong medicine. This does not just apply in psychiatry, it also applies in many other areas of medicine; for instance, in the diagnosis of airway respiratory infections, whether the patient has flu or a cold virus, or a bacterial infection; or in pain, to differentiate between headache and a migraine.

Diagnosis is a central part of the emerging electronic healthcare record landscape, where a patient's condition needs to be defined according to a discrete list of possibilities. Electronic health record adoption has increased from 10% to approximately 40% from 2009 to 2013 [94]. Medical data have a marked degree of inherent variability, uncertainty and inaccuracy, some of which is due to use of language, some due to the way that medical professionals reach diagnoses or select what to record and some due to the variability in clinical terms used by different healthcare disciplines [95]. This variability cannot be eliminated, but it needs to be recognised and controlled in order to make efficient and well-informed healthcare management decisions, and procedures to make this happen are underway right now everywhere that electronic health records are being implemented.

In the United Kingdom, the GMC guidelines say that once a diagnosis is reached, it should be communicated to the patient if they need it or if they do not need it but want it anyway. It would seem only logical that a diagnosis would necessarily pre-cede a prescription, so it would be a *sine qua non* of an off-label prescription (there may be some circumstances where a diagnosis is uncertain, but in those cases, the reason for the prescription should be given). I also see this as part of the process of informed consent and helpful for a patient's self-determination and treatment. There may be some patients who do not want to know their ailment, but I would think they would be the rare exception.

One way of conveying the diagnosis would be to write it on the prescription (or on a detachable part of the prescription if the patient wants to keep it private). A written diagnosis is also very helpful should the patient want, for example, to follow up with their own Internet research. There would be advantages to the control of healthcare costs in doing this, because the pharmacist would then know if the prescription was reimbursable in situations where reimbursement is dependent on indication (see p. 113). Nevertheless, at present, both in the United Kingdom and in the United States, the diagnosis is not written on the prescription in the majority of cases. I would argue, and strongly, that this needs to change. Currently, there is very limited dissemination of information to the patient on what is being used for what. I will deal with this in more detail in Chapter 8.

Up to now, I have focussed primarily on the nature of the medicine and how it is prescribed to the patient by the clinician. I have mentioned that an estimated 2 billion prescriptions are written off-label every year. Some of that is down to the ingenuity of the doctor in finding analogous clinical situations to the ones incorpo-rated in the medicine's label; but an awful lot is down to the way the medicines are marketed.

It is now time to deal with the way pharmaceutical companies maximise their commercial opportunity, find their way through and around the regulatory barriers, and in my view, game the system. Doctors may write the off-label prescription, but industry tells them what to write (or tries to).

CHAPTER 4

Gaming the system: the role of the pharmaceutical industry

Industry lobbyists like to portray pharmaceutical innovation as being driven by thousands of worker bees all in white coats, inventing life-saving medicines; industry detractors like to portray the enterprise as marketing driven, operated by greedy, sinister corporate executives with little care for patients. Both sides tend to represent the issues in black and white, but the truth lies somewhere in between, and the area of greatest greyness is to be found in off-label medicine. For it is here that the evidence is weakest, the safety is least and the profits most unjustifiable.

Pharmaceutical companies are not entirely evil, but neither are they charities. In a capitalist world, their primary concern is to return money to shareholders. They are intensively regulated by law and state institutions, and only secondarily by ethical restraints. If they can simultaneously make profits and cure sickness, all boats will rise: both investors and patients will benefit. In fact, figures suggest that there has indeed been a substantial benefit from pharmaceutical innovation in terms of patient well-being: in the last century, the world witnessed an unprecedented increase in life expectancy (see Figure 4.1), 40% of which in recent decades was attributed to pharmaceutical innovation in the form of new drugs; this is actually a remarkably high proportion [96]. Most of the increase is due to better control of cardiovascular illnesses, such as hypertension and stroke, but pharmaceuticals have also had a considerable impact on other diseases. Perhaps nowhere is this more clearly demonstrated than in HIV/AIDS, where 30 new medicines have been approved in the past three decades and have turned what was so recently a death sentence into a treatable, chronic condition. As well as saving and extending lives, pharmaceuticals have also improved their quality, such as the disability of multiple sclerosis or the debilitating nausea of cancer chemotherapy. There is always more to be done, and looking ahead, there are great expectations for the advent of better drugs for cancer and hepatitis: in fact, during the writing of this book, the advent of Sovaldi™ has come to pass, a drug that has been found to offer a cure for around 90% of patients with hepatitis C virus. As far as cancer is concerned, we shall see.

However, where there is greyness, moral rectitude comes a poor second to commercial success. In the murky area of off-label medicine, pharmaceutical companies have found a clever way of increasing sales at minimal risk and using the medical profession to help them in that endeavour. In this chapter, I will show how companies use the ability (and ambition) of the medical profession to prescribe freely, to leverage huge commercial advantage, and how the present situation facilitates bias over balance in the information available to support the real-world use of medicines, quite different from the ways sanctioned by the regulatory agencies.

Off-label Prescribing – Justifying Unapproved Medicine, First Edition. David Cavalla.
© 2015 John Wiley & Sons, Ltd. Published 2015 by John Wiley & Sons, Ltd.

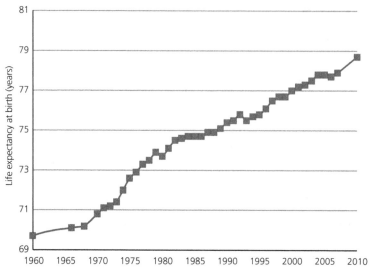

Figure 4.1 Life expectancy in the United States. Data from National Center for Health Statistics, National Vital Statistics Reports (http://www.cdc.gov/nchs).

Normal drug development and drug repurposing development

The regulatory approval process for a pharmaceutical product based on a new chemical entity takes 10–15 years and costs $1.8 billion per drug [97]. Actually, there is some debate about this figure, with those on the industry side of the fence adding in all the costs of failed developments and the 'cost of capital' related to investment returns they would have made on the failed development money; meanwhile, the industry detractors claim that the actual cost of a regulatory approval is manifoldly less and tend to assume (with 20:20 hindsight) that the failures could have been avoided. One simple way of getting a number that appears reasonable is to take the overall expenditure on pharmaceutical research and development (R&D) and divide it by the number of products that are approved each year. According to this method, the cost per approved new drug is actually staggering. The average drug developed by a major pharmaceutical company costs at least $4 billion, and it can be as much as $11 billion [98].

R&D costs for new medicines have risen enormously in recent years. This is because of a number of factors, such as that the diseases remaining to be treated are more difficult to address, that clinical trials to prove efficacy are increasing in size and scope, that high levels of safety are an increasing requirement for the drugs we take and that regulatory demands are increasing. The current estimates of R&D cost are now so huge that almost all the big companies have seen widespread reorganisations and restructuring to ameliorate these costs.[1] If it really was so easy and straightforward, and cost

---

[1]In fact, the scale of the retrenchment in the United Kingdom has been extraordinary. A report from 1996 on the level of neuroscience drug discovery in the United Kingdom identified research and development centres from the following companies: Abbott, Boots, Eisai, Fujisawa, Glaxo Wellcome, Janssen, Lilly, Merck, Nycomed, Organon, Otsuka, Pfizer, Reckitt & Colman, SmithKline Beecham, 3M, Xenova and Zeneca. Nowadays, all but three (GSK, Pfizer (Neusentis) and Lilly) have disappeared. Admittedly, there have been new companies established, but these are typically small biotech or virtual organisations that outsource much of their work and do not come anywhere close to the scale of the large companies that have disappeared.

as little as the most shrill industry objectors maintain, it is difficult to see why some of these smart people would not have done it. So, for the sake of argument, I am going to take it as read that getting a new pharmaceutical medicine to market is extremely difficult, risky, costly and time-consuming.

The conventional drug discovery process first requires testing candidate drugs against a biochemical target through large-scale biological screens, using chemical libraries and medicinal chemical variation. As the discovery process advances, the testing increases in sophistication: a lead compound is defined and tested in animals for toxicity. After this, in the United States, the innovator files an application with the FDA as an investigational new drug to begin trials on humans. A similar process occurs in Europe, involving the EMA, with offices in London's new dockland area. Three phases of human testing are required: preliminary testing in volunteers (Phase I), testing a small target population of patients who have the disease or condition under study (Phase II) and large-scale, randomised double-blind trials also in patients (Phase III). During all three phases, regulations require researchers to obtain the informed, written consent of participants; this is part of the ethical review and approval process which applies to all clinical trials. Once the required testing is completed, the manufacturer submits a new drug application (NDA) to the FDA or marketing authorisation application (MAA) to the EMA. In the past 20 years, the regulatory processing time required to obtain FDA approval of an NDA has dropped from (about) 22 months to (about) 12 months for most drugs and less than 6 months for drugs that treat life-threatening conditions.

'Attrition' is the word for abandoning R&D candidates at some stage before they obtain a licence to be sold, possibly because they are unsafe or do not work in clinical trials: it is a bug bear in the pharmaceutical innovation process that multiplies costs many times. It is a particular problem if the failures occur late on: Phase III developmental failures are enormously costly and far too common. When they happen, as with the spectacular failures of Alzheimer candidate bapineuzumab in 2012 or Pfizer's cholesterol drug torcetrapib in 2006 (among many others), the costs to the companies involved amounted to hundreds of millions of dollars. Executives scrabble for straws of hope when reporting to their shareholders' financial losses for the unfortunate investment, but mostly, these straws turn out to be worth nothing at all. Many observers claim with hindsight that the failures could have been predicted; some even do so meekly in advance and then crow much louder after the event. Alternative strategies have been proposed, with financial models attached. The Paul et al. reference upon which the figure of $1.8 billion was based, posits a method to reduce costs by failing developmental projects more quickly. Thus, risk is identified earlier, late-stage trial failures are less common, and overall costs are reduced, even though the attrition rate as a whole is not. This tells us something about the intractability of attritional failure in the pharmaceutical industry. In other words, rather than try to ameliorate the attrition, let's just accept it exists but try to find out about it earlier. Provided you don't end up throwing away too many good eggs along with the bad, it is a sensible approach, though not so easy in practice.

The long-standing figure for the number of products that enter early human trials (Phase I) relative to those that are ultimately approved is 10; 9 fail, and 1 succeeds to marketing approval. This is the figure used in the Paul et al. reference. In certain therapeutic areas, the number is higher, and in others, it is lower. This figure has remained relatively constant over the years. Figure 4.2 shows a simple graph of the percentage chance of a compound that enters into the first stage of human trials progressing to future stages of development. The data are based on a study of 4451 drugs from 2003 to 2011 and include probabilities of success for drugs in any eventual indication, not just the initial one chosen. This study found that the overall

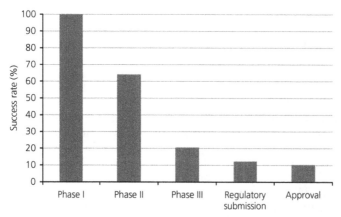

Figure 4.2 Percentage success of compounds entering first human studies (Phase I) that are successful in progressing to subsequent stages of development. Reprinted with permission from Ref. [99]. © Nature Publishing Group/Macmillan.

success rate for drugs moving from early-stage Phase I clinical trials to FDA approval is about 1 in 10 [99]. For difficult areas such as cancer, that drops to 1 in 20.

As any gourmand knows, one way of decreasing a restaurant bill is to skip the hors d'oeuvres and start with the entrée. So, obtaining a full regulatory approval for the examples taken from Chapter 2 is going to be significantly cheaper and quicker than full-scale development, purely because you can use the information acquired in the first development to allow you to abbreviate or avoid entirely some of the required studies. It is cheaper because the costs of the earlier stages of development can be avoided. But is it any less risky?

Well, to a certain extent, it is. The risk associated with intolerable side effects that might be picked up in Phase I, where a product's safety is first tested, is reduced. According to the graph, this amounts to 30–40%. However, it cannot be assumed that the dose required for the new use is the same as that for the original use. If this is the case, and the second indication requires a higher dose, the equation becomes rather different. One needs to include some Phase I risks associated with this higher risk. The true picture might then be rather worse, since in practice, it might be that the ideal dose is not identified until a dose-finding efficacy study is conducted in a Phase II trial.

The risk advantage, such as it is, becomes less clear in later stages of clinical trials. It is sometimes mistakenly assumed that the risk of a known drug failing for safety reasons in late-stage development is very small, on the basis that the safety of the product is already well established. In fact, there is precious little evidence for this, unless one is only considering a minor modification of the original therapeutic utility. In cases where a known drug is considered for a radically differ-ent new use, there are substantial risks, particularly if the new indication involves treatment with a new mechanism (for that indication) or requires better tolerated drugs than the original use provided.

Taking these factors together, it would seem that success rates in the development of secondary uses for existing drugs are somewhat greater than for new drugs, but developments of this kind are still much more likely than not to fail. A report by Thompson Reuters suggests the advantages are in shorter development times, reduced development costs and success rates of around 25% relative to the 10% or so for conventional new drugs [100].

The story of the secondary development of the drug Dimebon for Alzheimer's disease emphasises the kind of commercial costs and risks associated with this strategy. Dimebolin hydrochloride (Dimebon) was originally considered an antihistamine and had been approved for use in allergies in Russia since 1983. In the first decade of the new millennium, it was re-evaluated for its neuroprotective properties in Alzheimer's and Huntington's diseases. Dimebon has activity at two relevant and validated targets for Alzheimer's, so it was thought to be a good candidate for 'repurposing' in this challenging medical condition. The relevant biochemical activities were firstly as an inhibitor of acetylcholinesterase, which is the basis for rivastigmine (Exelon™), donepezil (Aricept™) and galantamine, which was mentioned in Chapter 2; and as a modulator of a subtype of the glutamate receptor (glutamate is a neurotransmitter), which is similar to memantine (Namenda™). All of these are marketed treatments for Alzheimer's disease; someone might assume that combining the mechanistic qualities of two classes of existing treatments would be a good approach to something that would be better. However, Dimebon had a range of other activities including histamine antagonism and other receptor activity which was not thought to be important for any benefit in Alzheimer's disease.

Following investigations in various animal models, Dimebon was given another name (latrepirdine) and studied by the biotechnology company Medivation in a large Phase II clinical trial in mild–moderate Alzheimer's disease. After 12 months of treatment, the results of the trial showed a significant improvement in cognition relative to placebo [101], and on the basis of this success, Medivation licensed the drug to Pfizer in a deal with a headline value of $725 million, excluding royalties. Indeed, to those not familiar with the vicissitudes of late-stage drug development, the early trials might indicate with almost complete certainty that the drug 'worked'. Pfizer apparently agreed and paid part of this licence fee ($225 million) as an upfront non-returnable commitment. Pfizer then paid further to conduct Phase III clinical trials, the costs of which are not revealed, but are certainly substantial. Unfortunately, these later trials did not fulfil the early promise, and the first Phase III results were negative – latrepirdine did not result in significant gains on any of the five efficacy endpoints versus placebo after 6 months of treatment, including 'mini-mental state examination' (MMSE), which is one of the standard measures of cognition. Not being willing to abandon a major investment, Pfizer continued with another three Phase III trials, all of which also failed. Still determined to identify some utility in a project in which they had placed so much hope (and money), further trials continued in Huntington's disease. In this too, latrepirdine demonstrated no beneficial effects in terms of either the MMSE or a composite assessment of the clinician plus caregiver. Finally, in 2012, Pfizer and Medivation both declared the development of latrepirdine to be abandoned.

Another example, this time demonstrating the impact on a small biotechnology company rather than a large multinational pharmaceutical company, involves a prospective new use for the selective serotonin uptake inhibitor, fluoxetine, better known by its trade name Prozac™. This drug has been used for decades as an antidepressant and for the treatment of generalised anxiety disorder and obsessive compulsive disorder. There was also evidence for the clinical benefit in a number of other indications, including autism. A series of clinical trials of fluoxetine in childhood and adult autism were carried out by Professor Hollander at the Mount Sinai School of Medicine in New York around the turn of the last century.

The results were taken forwards by the biotechnology company Neuropharm Limited, which was interested in the product's commercial position as an orphan drug (see p. 56) and in developing it as a new rapidly dissolving formulation which would have advantages to both the carer and to the patient. Specifically, the rapid disappearance of the product in the patient's mouth could help the carer to make sure the patient had taken the dose. Neuropharm licensed the rights to the programme in 2006 from Prof Hollander, feeling fairly sure that the Phase II trials would be replicated in their Phase III programme and that further hurdles to approval would be minimal, given the drug's well-known and accepted place in the pharmacopoeia.

Neuropharm launched on the UK Alternative Investment Market in 2007, raising £20 million from investors to pursue the necessary late-stage development. However, the trials did not succeed as planned, and the product failed to demonstrate a significant reduction in 'repetitive behaviour' in 158 autistic patients treated with fluoxetine over a 14-week period compared to placebo. The results were described as unexpected and disappointing by its chief executive in 2009, and the clinical programme was ultimately abandoned. The company filed for voluntary liquidation in 2010.

In addition to the risk of R&D failure and the enormous associated costs, another less well-appreciated fact is that even for those products that reach the market, a pot of gold rarely awaits. In fact, a substantial majority of marketed products fail to recoup their R&D investment. How can this happen? This reason is because the forecast for the product does not match reality; in fact, the majority of consensus market forecasts are off by more than 40% [102]. The innovator company may therefore commit to a development programme that over-promises; however, following its completion, the money has been spent, and it still makes sense to market a commercially unremarkable product. Of course, there are equally plenty of examples of mega blockbusters whose success is under-forecasted. It is these successes, viewed with hindsight, that flavour our view of an industry flush with unwarranted profits.

Provenge™ (sipuleucel-T) is a good example of marketing forecasts that over-promise. The product is a therapeutic cancer vaccine for hormone-resistant prostate cancer. It must be prepared from a patient's own cancer tissue, making the product specific for the patient's disease and body. The process involves taking a patient's white blood cells, incubating them with a protein specifically from prostate cells and generating a product which can induce an immune reaction when injected back into the patient. According to the theory, this immune reaction then permits patients so treated to destroy their own cancers. Thought by some to be a holy grail of cancer therapy, it was priced at $93 000 per patient per year, on the basis that it was the first cancer treatment that was a true cancer vaccine.

In December 2010 (8 months after FDA regulatory approval), one analyst said [103]:

> Should Provenge prove to be as successful as many in the biotech industry think it will be, peak sales of $2.5 billion is [sic] possible. This is partly because competition is likely to be minimal since barriers to entry are very high due to the long and arduous developmental cycle a similar product must endure to reach the market.

However, results have disappointed, generating $325 million in Provenge sales during 2012. Dendreon's forecasts are for sales to remain broadly flat during 2013: that is, roughly one-eighth of the 2010 forecast.

Now, imagine, if you will (and without any particular sympathy), the position of a hard-pressed pharmaceutical executive with an underperforming franchise. How can sales be increased quickly and without incurring the risk of a traditional developmental strategy? One way to address these problems might be to find new uses for existing products and find doctors willing to prescribe them off-label. The strategy would need to be based on generating some initial evidence of efficacy, but this might be done with a much smaller set of trials than are required for registration. The great advantage of this is that the dangers of developmental attrition could be largely avoided; provided a physician can be convinced to prescribe on the basis of some early evidence, there would be no need to undergo the substantially more stringent process of full regulatory approval. Secondly, the commercial returns can be achieved much more rapidly and cheaply, without the time and cost required for regulatory approval. And thirdly, even if the initial trials prove unsuccessful, the primary sales franchise would be undamaged. At least that's the theory.

Gaming the system

On the face of it, this strategy looks like a very good, almost one-way bet. Effectively, what is happening is that the pharmaceutical company is replacing the well-known difficulty of persuading the regulatory authorities with that of persuading a pre-scriber. That's why I call it gaming the system. It doesn't come without any risks or consequences, however. Although off-label use is usually advantageous for com-panies, it occasionally works against company goals. Before laying out how it can be done successfully, here's a couple of examples of what can go wrong.

Back in 2002, the biotech company InterMune was attempting to increase the market for Actimmune (Interferon gamma-1b). Actimmune had been approved by the FDA for chronic granulomatous disease, a rare genetic condition that affects the immune system, but InterMune were conducting an additional clinical programme designed to show efficacy in the treatment of idiopathic pulmonary fibrosis (IPF). IPF is a very serious lung disease, which is much more common than the labelled indica-tion (though still rare), and it was expected that 90% or more of Actimmune's sales were to come from the new use, regardless of whether the FDA approved it for such use. In other words, the company anticipated that off-label prescriptions in IPF would drive commercial success for the product regardless of the regulatory status. InterMune's investors, and in particular its CEO, Dr Scott Harkonen, awaited impa-tiently the results of a Phase III clinical trial to determine the efficacy of Actimmune in IPF.

However, when they came out, it appeared there were no statistically significant results for any of the trial's pre-defined clinical endpoints; despite this apparent failure, Dr Harkonen requested a series of retrospective analyses, one of which showed that only 4.8% of trial participants with mild to moderate IPF that received Actimmune died during the study, while 16.4% of such participants receiving placebo died, representing a 70% reduction in mortality. The difference was highly significant, but based on a retrospective analysis. We have dealt with the problem of retrospec-tive analyses previously and pointed out the analogy with 'picking your pocket' in the game of pool (Chapter 2). Suffice it to say that from the regulatory point of view (who, after all, were the key people this trial had to convince), it was only the prospectively defined endpoints which were going to be considered, and the retro-spective analysis was moot.

Scott Harkonen wrote a press release announcing the results of the Phase III clinical trial, characterising it as having 'demonstrated a survival benefit of Actimmune in IPF' and that Actimmune 'reduces mortality by 70% in patients with mild to moderate IPF'. The FDA did not agree with this and determined that it was a false statement because it arose from a retrospective analysis; the company had to pay $36 million to settle a Justice Department suit related to off-label marketing, on the basis that the release illegally promoted off-label sales of Actimmune. Harkonen was pros-ecuted and ultimately sentenced to 6 months' house arrest. He was threatened with loss of his medical licence, but this was not carried out; in any case, it makes little difference since he is barred from working in any company that is reimbursed by the US government healthcare system.

The InterMune story is unfortunate for Harkonen and Intermune, a bonanza for some well-paid lawyers, yet a mere pinprick compared to the corporate mantrap that caught Merck in their attempt to enlarge the market for Vioxx™. This story encapsulates not only huge financial consequences but also substantial patient harm and a brouhaha that persists – with claims of conspiracy that can only be partially countered with a defence of chaos.

Vioxx was one of a new breed of painkillers to succeed the widely known non-steroidal anti-inflammatory drugs (NSAIDs) like aspirin, naproxen, diclofenac and ibuprofen. These latter four pharmaceuticals are of course commonly used for all kinds of pain relief, from headaches to arthritis, and some of them are readily purchased without a prescription at high-street pharmacies. They share a common mechanism, namely, inhibition of an enzyme called cyclo-oxygenase (COX), which governs the levels of the inflammatory prostaglandins. However, it has been known for many years that they also cause gastrointestinal bleeding (gastropathy) and kidney injury (nephropathy). In fact, the gastropathic side effects of the NSAIDs are the most widespread adverse effects of any drug and are estimated to cause more deaths than from many common diseases, such as asthma, multiple myeloma, cervical cancer or Hodgkin's disease.

In the 1980s, the COX enzyme was subtyped into COX-1 and COX-2 versions, and evidence arose to suggest that the gastropathic effects of the NSAIDs were due to COX-1 inhibition. On the other hand, COX-2 seemed to govern the analgesic effect. In a classic pharmaceutical company strategy, Merck (and many others) pursued the identification of a COX-2-selective compound that could alleviate pain but do so without the gastropathic side effects. In May 1999, that project culminated with the approval of Vioxx (rofecoxib) for the treatment of acute pain and symptoms of arthritis. At the time, the available evidence from internal examination with endoscopy tests showed a significantly lower risk of gastrointestinal ulcers from Vioxx in comparison to the non-selective COX-inhibitor ibuprofen.

However, along the way, COX-2 had become associated with other diseases such as cancer and Alzheimer's disease, and COX-2 inhibitors were posited as treatments for these conditions (in fact, the action of aspirin on colon cancer is thought to be via this pathway). In order to investigate the effects of Vioxx on the prevention of colorectal polyps, a precursor of colorectal cancer, Merck started the 3-year Adenomatous Polyp PRevention On Vioxx (APPROVe) study in 2001. Unfortunately, the interim safety results that were revealed during the course of the study forced it to be abandoned after 18 months. By this time, a doubling of the risk of adverse thrombotic cardiovascular events (including heart attack and stroke) was revealed for patients on Vioxx relative to those on placebo. By comparison, the original regulatory trials for this product included approximately 5000 patients on Vioxx for periods of up to a year and did not show an increased risk of heart attack or stroke. It seemed that by extending the period over which Vioxx was administered, in an effort to extend its market from arthritis to cancer, an unexpected toxicity had been revealed.

Worse, much worse was on the way. The safety concerns revealed by the APPROVe study did not just relate to use in cancer, they also impinged on the primary use in pain. Realising that the existing usage of Vioxx included arthritic pain patients on long-term treatment with the drug, Merck was forced in September 2004 to withdraw Vioxx from the market because of concerns about increased risk of heart attack and stroke associated with long-term, high-dosage use. By this time, an estimated 20 million Americans had taken the drug.

The company also had to face up to accusations that it had ignored safety concerns from an earlier study, which had compared Vioxx to an earlier generation NSAID, naproxen. This study, with the acronym VIGOr, was intended to show a reduced risk of gastropathy with Vioxx compared to naproxen [104]. As such, it was a marketing study, rather than a regulatory one. What Merck were trying to do was reveal a better side effect profile for Vioxx in the treatment of arthritis relative to a pre-existing treatment, naproxen. It involved longer exposure times for patients on the drug, looking at clinically relevant outcomes like gastrointestinal ulcers, and consequent bleeding or death (and anything else that might happen after longer drug treatment times). Regulatory agencies normally ask for safety and efficacy studies that compare a trial drug with respect to placebo, but comparative studies (obviously) look for the differences between two alternative therapies. This is fine when there is a recognised benchmark for the thing being compared, but in other respects can just pose more questions. By way of illustration, let's consider the outcome from a trial of apples and oranges in cancer. If we showed that the patients on apples had more cancers than those on oranges, we would not know if apples actually caused cancer, or if oranges were protective, or if both of these things were happening.

The VIGOr study was published in the *New England Journal of Medicine* in 2001 [104] and showed that Vioxx was safer for the gut than naproxen; but it also revealed a four to fivefold difference between the drugs in terms of the number of incidents of heart attacks, with Vioxx having the higher risk. As far as gastropathy was concerned, it was already known that naproxen caused gastric ulcers, but the cardiovascular situation was more complex, since some older NSAIDs, like aspirin, were known to be protective for heart attacks, whereas other NSAIDs did not produce this benefit. Could the difference between naproxen and Vioxx in terms of cardiovascular risk be attributed to a protective effect of naproxen or a dangerous property of Vioxx? Here, the salient weakness of a comparative trial, which did not include a placebo arm, was revealed.

Merck was keen to publicise a headline of reduced gastropathy with Vioxx, but explained the cardiovascular difference as due to naproxen's shielding effect, rather than due to a hazard of Vioxx. Independent observers were not so sure. The FDA, who had received the VIGOr results 6 months in advance of the New England scientific publication, raised a question and asked Merck to explain. Alarms too were raised by some influential academics, including the eminent cardiologist Steve Nissen, who published an analysis of the data in a 2001 issue of the *Journal of the American Medical Association*, cautioning about an increase in cardiovascular risk with Vioxx [105].

Despite the public availability of Nissen's study, and indeed the knowledge of these safety concerns by the FDA, the period from 2001 to 2004 became a crucial one for litigation by patients who had suffered heart attacks. At the beginning of this period, we had information from the VIGOr trial which showed a relative risk, but it was only at the end that we had information from the APPROVe trial, where placebo dosing established that Vioxx was cardiotoxic in an absolute sense, and the product was withdrawn. For the prosecution, Merck simply knew about the increased cardiovascular risk with Vioxx in 2001, but chose to deny it and continue to sell a poisonous drug until 2004; the FDA were ineffectual, despite drawing the issue to Merck's attention, and even the New England scientific journal had been complicit in helping to publicise the reduced gastropathic risk while failing to alert readers sufficiently to the cardiovascular risk. For Merck, the defence was based on a restatement that the enhanced risk of heart attacks was only revealed from their colon cancer trial after 18 months on rofecoxib, that they fulfilled their legal obligations of disclosure of information to the FDA, that the publication of the meta-analysis in

2001 had publicly revealed Nissen's view of the relative heart attack risk from the VIGOr study and that the remaining errors and failures were a result of chaos rather than conspiracy.

At this point, I should perhaps mention that despite the increased number of heart attacks, the VIGOr trial showed no overall difference between naproxen and Vioxx in terms of overall mortality. To a certain extent, this could be because the trial was not big enough nor long enough to reveal a statistical difference, but is also reflective of the fact that the two risk factors, due to cardiovascular effects and gastrointestinal effects, counteract one another. Anti-Vioxx campaigners have estimated that 88 000 Americans had heart attacks from taking Vioxx while it was on the market, of which 38 000 died; but at least 500 000 would have suffered, and 85 000 even died from gastric bleeding as a result of taking conventional NSAIDs during the same 5-year period (see p. 101) – apples and oranges indeed.

The legal battle around Vioxx was momentous. A deluge of legal suits enveloped the company in the years following the withdrawal of Vioxx from the market. In 2007, Merck announced the establishment of a $4.85 billion 'settlement fund' to end thousands of lawsuits, believed to be the largest drug settlement ever. In addition, it was estimated to have incurred $1.9 billion in legal costs related to the debacle. Without admitting fault, the settlement allowed Merck to avoid the personal-injury lawsuits of some 47 000 plaintiffs and about 265 potential class action cases filed by people or family members who claimed drug-induced injury. The huge size of the settlement needs to be put in the context of the huge sales of the drug, however; in the year before withdrawal, Merck reported sales of Vioxx amounting to $2.5 billion.

The Vioxx story dramatically highlights the dangers of formal clinical studies for a secondary indication in damaging the franchise associated with the core, original indication. But, believe me, for every Vioxx, there are many more examples where the strategy bears fruit. The advantage of off-label use is not just that sales can be achieved without the burden of difficult, regulatory studies for the add-on indication. It is that the commercial returns are so vast. The challenge for the pharmaceutical company is therefore in finding the most effective means of convincing physicians to write off-label prescriptions with the minimum downside risk to the product.

Orphan use

A particularly advantageous off-label strategy derives from products that are initially approved for rare diseases. Such products are able to justify very high prices initially because there may be no existing products for these conditions, and the company can argue that low sales volumes need to be compensated by high tariffs, in order to return the R&D investments. They can also benefit from government programmes that have been designed to offer commercial incentives to bring products to the market for such conditions, which were originally called 'orphan' diseases since they lacked drugs to care for them. Of course, prices are not lowered once the products become widely used for off-label indications that are drastically more common than the rare approved disease, nor are the government incentives returned. Let's now look in more detail at how this can be a very attractive strategy for the pharmaceutical innovator, yet one of dubious morality.

In the United States, Congress passed the Orphan Drug Act in 1983 to provide incentives for industry investment in treatments for rare conditions which, absent the incentives, were not attractive for pharmaceutical industry R&D investment. Orphan drug indications concern over 6000 diseases and span a very wide range of therapeutic areas, which are similar only insofar as they are rare; because the orphan diseases are so numerous, the total number of people affected is estimated at up to

20 million people in the United States and 30 million in Europe. And as we shall see, they are, in some cases, the foundation for products which command very high prices, large revenues and are widely used off-label.

In terms of new drugs on the market, the Orphan Drug Act has been a notable success. As a result of this act, over 300 drugs and biological products for rare diseases have been approved by the FDA since 1983 (compared with fewer than 10 such products in the decade prior to 1983). The US legislation enables fast-track regulatory approval, marketing protection, tax incentives and funding for clinical research in rare diseases. One thing it does not offer, though, is any relaxation in the central principle of proving adequate safety and efficacy in order to obtain regulatory approval: this element is sacrosanct. Once a drug is approved, a big attraction for an Orphan product is that a generic competitor in that indication cannot be approved for 7 years (normal patent protection may also apply). As a result of the success of the US act, similar legislation is now in place in Europe, Australia, Singapore and Japan, with each jurisdiction having a slightly different definition of an orphan indication. For instance, whereas in the United States an orphan designation may apply if the prevalence is less than 200 000 (approximately 6.5 in 10 000) and offer 7 years' market exclusivity, in Europe, the corresponding figures are a prevalence of 5 in 10 000 and 10 years' exclusivity. The key features of the various orphan drug acts around the world are shown in Table 4.1.

Table 4.1 Summary of orphan legislation around the developed world.

	United States	**Japan**	**Australia**	**European Union**
Date established	1983	1993	1997/8	2000
Legislation	US Orphan Drug Act modified the Federal Food, Drug and Cosmetic Act	Partial Amendments Law amended two previous laws	Additions made to the regulations to the Therapeutic Goods Act 1989	Regulation (EC) #141/2000
Designation requirement				
Prevalence	Less than 200 000 (6.5 per 10 000)	Less than 50 000 (4 per 10 000)	Less than 2000 (1.1 per 10 000)	Less than 5 per 10 000
Financial return on product	If costs cannot be recovered	No	If costs cannot be recovered	If costs cannot be recovered. Returns considered over a period of 7 years
Disease type	Rare only	Rare and serious disease, no other treatment available	Rare only. Normally for serious and life-threatening diseases, although not required by law	Life-threatening or chronic debilitating, no alternative treatment
Incentives				
Market exclusivity	7 years	Re-examination period extended from 4 to 10 years for orphan drugs	None	10 years
Fee waiver	Yes	No	Yes	At least partial
Grants	Clinical studies for pharma and academia	Clinical and non-clinical studies, pharma only eligible	None	
Tax credits	Yes	On request	On request	Yes

Despite the fact that orphan designation applies to rare diseases, and may drive the commercial viability of a new therapeutic for a particular disease, there are notable examples of much wider use which substantially broaden the market access for orphan drugs. For instance, in 1989, the FDA approved the use of epoetin alfa, an erythropoietin-stimulating agent (ESA), for patients with anaemia in end-stage kidney disease. Erythropoietin was first isolated by the researcher Eugene Goldwasser in 1972, and in the early 1980s, he approached what was then a start-up company in California, Amgen, who were looking for products to license. Amgen gave the product the trade name Epogen (later, in longer-acting form, Aranesp™), but co-marketed it also with Johnson & Johnson (J&J), who adopted the trade name Procrit.

In addition to use in kidney disease, it was also later approved for the treatment of anaemia associated with HIV. Both of these two indications were also approved with orphan designation. Yet despite the restriction to use in end-stage renal disease, the market very quickly expanded to include nearly all kidney dialysis patients, not just the roughly 16% who required blood transfusions. Fed by direct-to-consumer (DTC) television and print advertising campaigns claiming 'statistically significant improvements for … health, sex life, well-being, psychological effect, life satisfaction, and happiness', Epogen became a hugely profitable product. In addition to the initial approved uses, ESAs were subsequently approved for the treatment of anaemic patients scheduled to undergo elective, non-cardiac, non-vascular surgery, such as hip and knee surgery, and for the treatment of anaemia in patients with non-myeloid malignancies due to chemotherapy. It was through these non-orphan uses that Amgen derived much of its revenue. It was clear, however, that the ESAs were used even wider still, for anaemic cancer patients not undergoing chemotherapy. At the peak in 2006, sales of Epogen and Aranesp amounted to $5.3 billion [106].

During its commercial life, the size of average doses tripled, in accordance with Amgen's recommendations; from an initial regimen of 3500 units, the drug's label was changed to list a starting dose of 10 000 units, with a proportionately increased price. These increased doses were designed to do more to increase red blood cell count, which is conventionally indicated by the haemocrit level, the percentage of the volume of whole blood that is made up of red blood cells. Starting with an anaemic patient with a blood haemocrit of 25% or less, the higher doses were designed to raise this to 'normal' levels of 40%, whereas the lower-dose ESA regimen raised them to 30%. The higher doses started to raise safety concerns, since in addition to raising oxygen-carrying capacity, ESAs also made the blood thicker and more prone to clot. Thus, whereas early published clinical studies and meta-analyses appeared to show a benefit of ESAs on mortality, experience with higher doses of ESAs suggested they actually increased thromboembolic events and overall mortality in cancer patients, as well as higher levels of malignancy.

In response to the increasing safety concerns, in 2007, the FDA announced that Amgen and J&J would add new 'black box' warnings, the regulator's most serious type of warning. The warnings indicated the potential for ESAs to increase the risk of heart attacks and death, particularly in the higher doses recommended for the products. Since this time, sales of Epogen have levelled off, and sales of Aranesp have reduced to a total of around $3 billion in 2011. Even at this level, however, the ESAs are blockbuster products that do not need orphan designation for their commercial existence (such benefits are nevertheless retained).

There is also a seamier side to the use of ESAs, outside medicine altogether. In the 1980s, marathon runners, Nordic skiers and Dutch cyclists were gaining access to ESAs on the black market and using the drugs to boost their performance. From 1987 to 1990, about 18 young cyclists died under mysterious circumstances, including the 27-year-old Dutch cyclist Johannes Draaijer. In 1989, he suffered a heart attack, and although his sudden death was officially unexplained, his widow pinned it on erythropoietin. Then, at the 1998 Tour de France, police found hundreds of erythropoietin vials on members of the Festina team. Authorities raided cyclists' hotel rooms and found more vials on the TVM team's premises. But it was Lance Armstrong, who won the Tour de France a record seven consecutive times, more than any other sportsman, who made the ESAs infamous through the doping scandal that he finally admitted in 2013. In the previous June, the US Anti-Doping Agency (USADA) charged Armstrong with having used illicit performance-enhancing drugs and subsequently issued a lifetime ban from competition in all sports. In their report, the USADA concluded that Armstrong engaged in 'the most sophisticated, professionalized and successful doping program that sport has ever seen'. Armstrong was ignominiously stripped of all his Tour de France titles, and ESAs were his chosen agents of deception.

Erythropoietin has travelled a long way up from treating life-threatening rare anaemia to generally improving 'well-being and happiness', followed by a long fall into disrepute as a drug that can cause deaths or can be used to cheat in sport. It is also another example where off-label use and declining safety are tied to increased dose, a factor which we shall deal with in more detail in Chapter 5.

ESAs are not the only top-selling drugs with orphan status. The 100 top-selling pharmaceutical drugs in the United States by retail sales in 2009 included 12 approved for one or more orphan indications. Among those products, five were targeted for orphan diseases at the time of first FDA approval. For the other seven, the orphan indications were identified, and the benefit applied to the innovator company after the drug was already on the market.

A study by Jerry Avorn, professor of medicine at Harvard Medical School and a respected commentator in this field (and others), found that most of the commercial return for these valuable products was attributable to unapproved uses [107]. For instance, over the period to March 2005, the prescriptions of the lidocaine patch (Lidoderm™) grew rapidly, so that by the end of the period, around 85% of the uses were non-approved (see Figure 4.3), and this portion was growing at a faster rate than the approved orphan use.

The extent of off-label use of modafinil has been referred to before (see p. 17); in the form of Provigil™, Cephalon's trade name product, US sales increased from

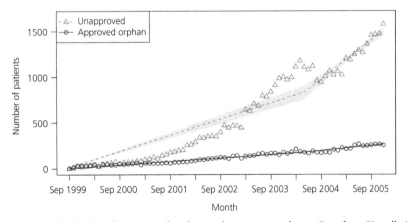

Figure 4.3 Use of Lidoderm for approved orphan and non-approved uses. Data from Kesselheim AS, et al. (2012) The prevalence and cost of unapproved uses of top-selling orphan drugs. PLoS ONE 7(2): e31894.

Table 4.2 Orphan indications for some top-selling drugs.

Drug name	Manufacturer	FDA-approved indication	Orphan designation	US sales ($ bn; 2011)
Lidocaine patch (Lidoderm)	Endo Pharmaceuticals	Painful hypersensitivity and chronic pain in post-herpetic neuralgia	Yes	1.19
Modafinil (Provigil)	Cephalon	Excessive daytime sleepiness in narcolepsy	Yes	1.11
		Shift-work sleep disorder	No	
		Adjunctive treatment of sleep apnoea	No	
Cinacalcet (Sensipar™)	Amgen	Hypercalcaemia in patients with parathyroid carcinoma	Yes	0.81
		Secondary hyperparathyroidism in patients with chronic kidney disease on dialysis	No	
Imatinib (Gleevec™ [Glivec™ in the United Kingdom])	Novartis	Chronic myelogenous leukaemia	Yes	1.47
		Gastrointestinal stromal tumour	Yes	
Glatiramer (Copaxone™)	Teva	Relapsing–remitting multiple sclerosis (RRMS)	Yes	2.86

Data from Kesselheim AS, et al. The prevalence and cost of unapproved uses of top-selling orphan drugs. PloS One 2012;7(2):e31894.

$25 million in 1999 to $475 million in 2005, to $800 million in 2007 and to $1.1 billion in 2011. The extent of off-label use has been put at 90% [55], which equates to over a billion dollars of 2011 sales in the United States alone. As shown in Table 4.2, there are two non-orphan approved uses for this drug; however, for the period to mid-2005, these contributed little to the use of modafinil. A further study of the off-label use of modafinil over the period from 2002 to 2009 found that prescriptions for patients with diagnoses of narcolepsy, shift-work sleep disorder and sleep apnoea grew only threefold, compared with an increase of 15-fold for off-label uses [55] (Figure 4.4). Meanwhile, patients with depression accounted for 18% of prescriptions, and those with multiple sclerosis accounted for 12%.

ESAs and modafinil are two of many similar agents that are used to treat unapproved conditions far more commonly than they are used to treat approved conditions, but the unapproved conditions are treatable without recourse to off-label medicine – there are licensed alternatives. Why do physicians prefer to use unapproved agents rather than approved ones? In these circumstances, it is actually quite difficult to defend the need for patients to be treated in this fashion. Unlike the situation where a patient is legitimately granted off-label access to a drug by a doctor who cannot find an approved agent, we now have drugs with expensive prices framed by their 'orphan' utility, actually being used off-label for much more common conditions, without regulatory approval. How has this come about?

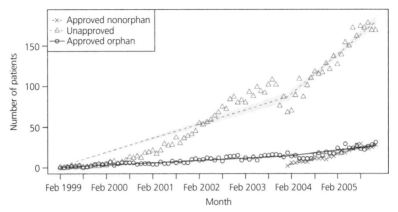

Figure 4.4 Use of Provigil for approved orphan, approved non-orphan and non-approved uses. From Ref. [107].

Pharmaceutical marketing

The pharmaceutical industry likes to portray itself as a research-driven enterprise that operates in the white heat of innovation. Unfortunately, this claim is efficiently deconstructed by the comparison of the huge costs of sales and marketing relative to those of R&D.

According to the industry lobby group Pharmaceutical Research and Manufacturers of America (PhRMA), research-based pharmaceutical companies in the United States spent $27.7 billion for all promotional activities compared to $29.6 billion on R&D in 2004. However, marketing costs are actually much higher than their lobbyists allow, according to a more detailed estimate from an analysis of the traditional gold standard source, IMS Health, together with figures from the market research company CAM [108]. IMS monitors 75% of prescription drug sales in over 100 countries and 90% of US prescription drug sales; its figures include amounts spent on visits by sales representatives (detailing), samples, DTC advertising and journal advertising. CAM, on the other hand, audits promotional activities of the pharmaceutical industry in 36 countries worldwide. The authors of this study give various reasons why they think promotional spend is higher than generally acknowledged, including a desire by the companies themselves to foster a better public image. The result of their detailed analysis for the year 2004 indicates a total of US$57.5 billion spent on promotional activities, more than double the figure given by the industry organisation PhRMA and approximately double that of the R&D spending.

We do not know how much is actually spent on that portion of promotional activity related to off-label medicines, since this particular and illegal category does not show up even in the figures available from IMS and CAM. To probe this in more detail, we can look at the cases brought by the regulatory authorities (mainly the FDA) against large pharmaceutical companies for off-label promotion. One of the first celebrated cases of this kind was brought against Parke-Davis/Warner-Lambert (which later in 1999 merged with Pfizer) in relation to gabapentin (Neurontin™). It was one of the first examples where a substantial fine was imposed for off-label marketing. As a result of the FDA fine, some things became public knowledge; one is a motivational speech by a Parke-Davis executive given to the company's marketing managers. At the time, Neurontin was approved for adjunctive treatment (i.e. as an 'add-on', rather than as an initial 'first-line' medication) for partial complex seizures at a dose of 1800 mg/day:

I want you out there every day selling Neurontin...We all know Neurontin's not growing for adjunctive therapy, besides that's not where the money is. Pain management, now that's money. Monotherapy [for epilepsy], that's money...We can't wait for [physicians] to ask, we need [to] get out there and tell them up front. Dinner programs, CME [Continuing Medical Education] programs, consultantships all work great but don't forget the one-on-one. That's where we need to be, holding their hand and whispering in their ear, Neurontin for pain, Neurontin for monotherapy, Neurontin for bipolar, Neurontin for everything. I don't want to see a single patient coming off Neurontin before they've been up to at least 4800 mg/day. I don't want to hear that safety crap either, have you tried Neurontin, every one of you should take one just to see there is nothing, it's a great drug.

Off-label promotion is illegal, but providing information to legitimate medical enquiries is not. So, 'cueing' the doctor to ask an off-label question, the pharmaceutical rep can then arrange for the company to send a packet of off-label information from the medical affairs office. The packet would typically contain company-approved reprints and a standardised letter, created by the medical information department, discussing research on the off-label use. These and other marketing strategies that were used in relation to the US promotion of gabapentin (Neurontin) were found in pharmaceutical industry documents as revealed by litigation and the resultant US congressional inquiry into the off-label case against Pfizer [109].

Consistent with the familiar pattern, the off-label component of the drug sales grew disproportionately compared to the sales of the on-label component. The sales by indication shown in the Figure 4.5 graph corresponds to a time period during

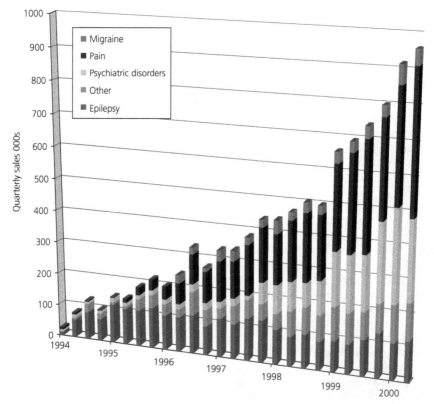

Figure 4.5 Neurontin (gabapentin) prescriptions, by indication. Graph drawn from data from Ref. [110]. (*See insert for color representation of the figure.*)

which the only FDA-approved use of gabapentin was for adjunctive treatment of epilepsy in adults older than 12 years. It is abundantly clear from this that the practice of off-label medicine provided enormous financial gains to Parke-Davis. The commercial advantages of this strategy are huge and are difficult to overstate.

Neurontin was promoted by activities that in other fields we might not typically recognise as promotional, such as education and research studies. But we must remember that the buying choice for a pharmaceutical is made by a highly discriminating and experienced group of medical professionals, so the banal ads that we would expect to see on television for soap powder are not likely to be effective. The selling tactics included 'independent' continuing medical education, 'peer-to-peer selling' by physician speakers, industry-funded clinical studies and publications in the medical literature. An example of the education component involved physician moderators directing audiences to discuss 'how Neurontin evolved into a first-line therapy option in your practice', when the label says it should be used as an adjunctive treatment. As another example, one medical education company used an unrestricted grant from the drug company to organise a series of study programmes on the use of anti-epileptic agents for the treatment of chronic pain. Chronic pain was a huge commercial opportunity, but not an approved use of gabapentin. Another medical education company received a grant from Parke-Davis to organise consultant meetings to train speakers to deliver lectures on anticonvulsant use in non-epileptic conditions. Again, gabapentin was labelled for epilepsy, *only*. As regards research studies, Parke-Davis employed a publication strategy focussed primarily on expanding gabapentin use in neuropathic pain and bipolar disorders, for which detailed marketing forecasts projected the greatest revenue potential. The strategy for these studies was not towards convincing a regulatory agency, but towards a much lower level of proof. This is not an exhaustive list of all the activities, just some of them. Ultimately, the drug company admitted guilt in connection to charges that during the 1990s it violated federal regulations by promoting the drug for pain, psychiatric conditions, migraine and other unapproved uses (see also Chapter 7).

Marketing tactics also include scientific papers that have been often written by company-paid ghost writers but bear the name of a medical school faculty member who was paid generously for the use of their name and then presented as part of 'educational' programmes. These tactics are much more effective in off-label medicine than efforts directed towards a new indication that undergoes regulatory scrutiny. Regulators have the time and resources, expertise and scepticism to analyse the submissions in detail, whereas a medical professional lacks the same resources and, of course, the time for an equally thorough analysis.

Whistleblower reports also show that marketing activities of this kind are rife. Gwen Olsen, a former pharmaceutical industry employee, tells the story of her time as a sales rep in Texas. She was responsible for the drug Haldol™ (haloperidol), a first-generation antipsychotic from J&J approved for schizophrenia and also used to control motor and speech tics in people with Tourette's syndrome; however, she was stationed near Corpus Christi, in an area with few antipsychotic specialists. Faced with a poor sales performance for 1 year, she was advised by her district sales manager to visit the local nursing homes, where she promoted the drug for disruptive behavioural in demented patients. She has since become a vocal critic of pharmaceutical company marketing practices, giving speeches and interviews on the subject, some of which are now posted on the Internet as Youtube videos; although she worked at a large pharmaceutical company, she also has a personal connection to what she felt was wrongful use of painkillers, antidepressants and antipsychotics leading to the suicide of her niece.

These are poor testimonies to the probity of the pharmaceutical industry. As detailed in Chapter 7, there have been substantial fines imposed on the industry from regulator-led inquiries and resultant litigation. There are some indications that public opprobrium may be leading some companies to reconsider this strategy, but other signs that the behaviour continues, with fines just a 'cost of business', and promotional activity considered a right of commercial free speech. Companies like GSK are currently attempting to overturn years of poor public relations, with an announcement in 2014 that it will no longer pay external physicians to market its products; but not all companies are adopting a similar attitude, and it remains to be seen how much of substance lies behind the reassuring words. Off-label medicine remains an attractive commercial honeypot and one that cannot be easily ignored. It is unlikely that public censure alone will cause a repudiation of the practice, with further concern that legal retribution is inadequate protection against a broken system. Let's not forget that generic products, too, which are not marketed, are also substrates for off-label use. A composite solution needs to be more holistic.

Expanding uses for non-pharmaceuticals

It is interesting to compare the situation for off-label use of pharmaceuticals with that of substances like marijuana which have a complicated quasi-legal position in many countries. For instance, Nabiximols (trade name Sativex) is a cannabinoid mouth spray developed by the UK company GW Pharmaceuticals for the treatment of spasticity in multiple sclerosis patients and approved in many countries in Europe as well as Canada. It is also being developed for cancer and various other forms of neuropathic pain. It is not FDA approved at present, and the difficulty in the United States is that marijuana is categorised as a Schedule I drug 'with no medical use'. As a result, it is regulated by the Drug Enforcement Agency rather than the FDA. Despite the prohibition at the federal level, around 20 states have individually approved 'medical marijuana', and when combined with the traditional laxity for off-label usage, this has led to the widespread use of this substance for a huge range of conditions.

There is evidence, though not to regulatory standards, for the utility of marijuana in the treatment of chemotherapy-induced nausea and vomiting, anorexia in AIDS, various neuropathies, fibromyalgia and the pain of rheumatoid arthritis and Tourette's syndrome. There is weaker evidence for its effects in Alzheimer's disease, glaucoma, overactive bladder, epilepsy, post-traumatic stress disorder, anxiety, depression and type I diabetes. It has been possible since the passage of Proposition 215 in 1996 to obtain a medical prescription in California for medical marijuana for any serious condition for which marijuana provides relief. It has even been prescribed for the treatment of insomnia.

By the extensive use of off-label marketing practices in support of its commercial ambitions, the industry runs the risk that the use of pharmaceuticals for poorly evidenced therapeutic purposes will be drawn in direct parallel to the use of medical marijuana for whatever you want. In case you think this is a strong statement, please see the section on Xyrem™ (sodium oxybate), another scheduled drug with a string of potential pharmaceutical uses, as well as illegitimate ones (p. 139).

DTC advertising

A lot of the attention given to the issue of off-label medicines is focussed in the United States. Partly, this is because there is a more permissive regime for this from the point of view of the medical associations (see Chapter 3) and a statutory 'hands-off' approach to the control of medical practice from the regulatory authorities (see Chapter 7). But another contributory reason could be the role of DTC advertising for prescription pharmaceuticals. DTC advertising is a familiar practice in the United States, where it may appear in television, newspaper, magazine and radio formats and in other mass or social media. It is banned in Europe and all other countries around the world except New Zealand; in these countries, consumer advertising is restricted to over-the-counter, non-prescription medicines.

The US stance on this issue is framed by its constitutional position on free speech, enshrined in the First Amendment. The FDA is responsible for regulating DTC advertising in the United States, despite having stated misgivings that 'direct to the public prescription advertising was not in the public interest'. Ultimately, the constitutional position trumped the regulatory one, and after a period of sporadic DTC advertising followed by a moratorium, the FDA effectively gave the green light to the practice in 1997 when it released its draft guidance on the use of broadcast advertisements. This guidance drastically changed the way that pharmaceutical companies were allowed to advertise, permitting advertisements to be accompanied with much abbreviated risk statements. Since then, spending on this promotional form has grown from $220 million to $3.4 billion in 2012.

One consequence of DTC advertising is the greater pressure on the prescriber, who is now squeezed on one hand by pharmaceutical reps and on the other by the patient, who may ask for a product by name, citing a TV commercial or magazine advertisement. A patient is much less likely to differentiate his or her condition from the precise wording contained in a pharmaceutical label, and much more likely to conflate his experiences with similar ones in an ad. DTC advertising was at least partly responsible for the rise and rise in sales of ESAs, since they talked about the effect of the products on 'fatigue', an incredibly common symptom of a huge range of diseases and non-diseases.

Although hard evidence for the causal association between DTC advertising and off-label prevalence is hard to come by, some health professionals such as psychiatrist David J. Muzina, from the pharmacy benefit manager Medco, believe that DTC advertising has helped fuel the rising use of antipsychotics [111]. The role of DTC advertising in 'medicalisation', in other words persuading non-patients to view their condition as a disease, thereby increasing unwarranted diagnoses, has been posited. Similarly, the language used in DTC advertisements has been found to contain incomplete syllogisms, leading audiences to make an emotional assumption that their symptoms can be medicated. There is also evidence that prescriber choices can be influenced by patient preferences, mediated by DTC advertising [112]. However, set against these facets, FDA regulation prohibits advertisements directed towards patient groups or indications outside the label, so pharmaceutical companies would need to cover themselves with at least plausible deniability in causing any increase in off-label medicine. It is therefore difficult to conclude that DTC advertising promotes off-label use, but if it promotes overall use of a drug, it is also likely indirectly to promote the absolute numbers, if not the proportion, of off-label prescriptions.

Patents and genericisation

A discussion of the behaviour of pharmaceutical companies is incomplete without mentioning patents, along with other forms of data protection which are offered by governments from most developed countries. Patents are a negative right, in other words they give the patent holder the right to prevent a competitor impinging on the exclusive rights held within the patent; this is a subtly different thing from the right to do something. Without some form of protection of this kind, there is no viable commercial model for investment in pharmaceutical R&D. Patents prevent price competition during the period of their validity, beyond which generic companies enter the market. The price difference between patented and generic medicines goes to support the claim that the branded section of the market holds patients hostage to overpriced therapeutics. However, generic companies feed on the products that were first introduced by innovators. Today's branded medicine becomes tomorrow's generic, and the interval between the two variants is actually quite short; paradoxically, the appetite for generics is spurred by high-priced branded products, and the so-called patent cliff in which a whole generation of branded medicines lost patent protection since 2010 has resulted in a feeding frenzy of generic competition in recent years. Generic companies have to make a profit, too, and because they work on non-patented products, they have no reason to invest in research or clinical trials; anything they do along these lines is shared out among all their competitors and effectively wasted.

The world patent system divides patent rights into different countries, all with different rules. However, there is increasing homogenisation of those rules around a 20-year patent life, starting from the date of filing. In practice, companies file patents at the beginning of a project, and the R&D and subsequent regulatory review process take a substantial bite out of the available period of monopoly. As a consequence, legislation was put in place decades ago such that patent offices in most developed countries could offer various forms of supplementary patent protection for marketed products for up to around 5 years. In addition, drugs are often protected by more than just one patent. Famously, the period of exclusivity for ranitidine, the anti-ulcer product, was extended by a follow-on patent, covering a specific crystalline form of the world's first billion-dollar drug. With all this complexity, the eventual period of exclusivity can be difficult to predict and may vary considerably. However, it is also possible to look at the approval history of all drugs and their generic equivalents and, by looking at the gap between the two events, calculate the effective commercial monopoly time. The FDA compiles the so-called Orange Book with this information available online (http://www.accessdata.fda. gov/scripts/cder/ob/default.cfm). Looking back at the period from 1982 to 2013, I calculated the average period of exclusivity to be around 12.5 years. There is a slight trend downwards in the exclusivity period over time, in agreement with others like DiMasi and Paquette [113], who have posited this trend but not provided hard evidence for it.

There are plenty of other things that could be written about patents, but they would be irrelevant for this book. As far as secondary uses for drugs are concerned, there may be two patents covering these forms of development. The first concerns the original product, a so-called composition of matter form of protection, that is also likely to claim the product's original intended use. Then there is the patent covering the specific secondary use. As mentioned earlier, patents are a negative right, so the holder of the first patent can prevent the holder of the second (if different) from developing the secondary use while the original patent still holds. Once the original patent has expired, the second patent holder then can operate his right and prevent everyone else competing, but only in regard to the secondary use. If both the composition of matter patent holder and the method of use patent holder are the same person, there is a period of patent coverage by the same person or company across the two patent periods.

Upon genericisation, prices drop substantially as competition gets underway: the magnitude of the drop depends on the number of generic competitors, but can be

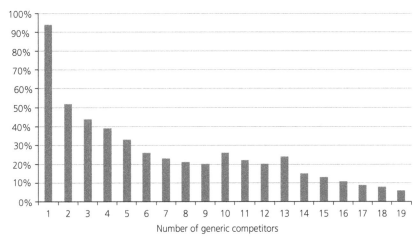

Figure 4.6 Average relative price of generic to brand. FDA analysis of retail sales data from IMS Health, IMS National Sales Perspective (TM), 1999–2004, extracted February 2005 (http://aspe.hhs.gov/sp/reports/2010/GenericDrugs/ib.shtml).

over 90% (see Figure 4.6). The information database on a drug that has been deposited by the originator with the regulatory authorities is then used as a point of reference for the newer generic products. They are approved on the basis that they can effectively substitute for the original products. Generic products are much less actively marketed than the branded products, if at all; the lack of marketing helps to keep prices low, and helps to keep off-label practice from rising further, but does not reduce this component from pre-genericisation levels. The generic companies have practically no obligation or commercial incentive to provide any further information on the efficacy of the products in situations that might pose as 'off-label'. But there is still a problem for patients if significant levels of off-label prescriptions have been written during the period of originator exclusivity, since from the aforementioned modafinil example, the prescription behaviour is unlikely to change once the product becomes generic. In fact, with a price manifold less, one could expect the levels of both on-label and off-label prescriptions to surge once genericisation has occurred.

The second problem with this situation is that there are usually no incentives for legitimate development of secondary uses of existing products once they become generic. The examples in Chapter 2 were properly developed because they had some form of patent protection available to them, as well as a differentiating factor. For example, Propecia™, the hair loss product, was covered by a method of use patent, which protected the product from competition long after the patent on its active ingredient (finasteride) had expired; in addition, it was differentiated from the generic finasteride for prostate hypertrophy by virtue of having a fivefold lower dose of active ingredient in the tablets. To do the same for drugs after they have become generic, one needs to patent them in some way (such as for a new use) and, in addition, find a different dose, or a different presentation, or some other feature that differentiates the generic product from the secondary product one is seeking to develop. If this kind of thing can be done, it can become a very attractive pharmaceutical strategy [114], but there are many examples where product development does not happen and the evidence base is left inchoate because of the lack of commercial incentive. Rather than adopt this route, pharmaceutical companies generally prefer to develop completely new drugs with a good, independent patent position, where existing ones would work very well. Ultimately, medicine is the poorer for this, and patients are left with fewer, more expensive options that take longer to come to the market.

The third problem about the sensitivity of the market to patent expiry dates is that products that can command greater periods of monopoly are more likely to be developed, regardless of the merits of the medicine. Although the average monopoly period for pharmaceuticals is 12.5 years, a much longer period applies for biologic drugs. To give an example, the ESAs were first introduced in 1989; yet, 25 years on, the genericisation of this market proceeds very slowly; the so-called biosimilar approval process is dogged by questions of what exactly can be substituted for what, and branded products remain dominant in this area, even at very high prices (its annual cost to each US patient was $8447 in 2009). Competitor products need to be tested for long periods to demonstrate their individual safety and efficacy profiles, whereas generic pharmaceuticals based on synthetic chemistry can merely undergo a series of equivalence measures. Similar stories apply almost everywhere in regard to biological pharmaceuticals. The rheumatoid arthritis products Enbrel and inflix-imab reign supreme in their areas, despite also having been introduced in the 1990s and patents having long since expired. Again, these medicines are also expensive, a feature which is partly attributable to their high costs of manufacture.

The strategy of developing biological drugs therefore has great attractions for pharmaceutical companies, including high prices and long product innovation cycles. However, there is more: these products generally have to be injected rather than taken orally, so they are less convenient for the patient. In addition, the active prin-ciples are large molecules which do not normally cross biological membranes, so making it difficult to target biological pathways that lie in the brain (since the brain is protected by a barrier that is impermeable to biological drugs). The consequence of this is that R&D effort is incentivised towards products based on biologics, with consequent therapeutic attention away from areas that are poorly addressed by these agents, such as neurological disease.

Conclusion

In this chapter, I have outlined how pharmaceuticals are developed and the risks involved. Finding means of expanding the market for a product once developed via off-label means is highly attractive. Examples such as gabapentin and modafinil show how the off-label component of the market can dwarf the on-label portion. The commercialisation of a product for off-label use does not require the same investment of time or money or expose the manufacturer to the same level of risk as new product development. Nor does it deliver the same degree of patient efficacy and safety as if it were sanctioned by regulatory approval. Indeed, it seeks to bypass the regulatory system, and when used for a majority of prescription activity, it makes regulation irrelevant for the bulk of the commercial return.

Orphan uses are therapeutic areas where additional government incentives have been put in place to enhance the commercial incentives for products for rare dis-eases; prices of orphan products are generally high because the scale of the market opportunity is thought to be poor. However, off-label use of orphan products can give rise to blockbuster products. There is an alarmingly small gap between some legitimate pharmaceuticals and illegitimate street drugs.

Pharmaceutical marketing practices of this kind are examples of 'gaming the system', and regulatory agencies, particularly in the United States, have been spurred into action to try to control it. (Actually, this is a fairly mild form of disapprobation; when you have read further in Chapter 7, you may wonder if 'criminal', 'corrupt' and 'abusive' are more apt descriptors.) Marketing practices can be very effective at driving off-label prescriptions, since it is easier to persuade a physician than the regulators. Once a product has become generic, the marketing practices may have calmed, but the prescribing practices are left in place, with the potential for patient harm. No further data to support these prescriptions are likely to be produced by manufacturers once the product has become generic.

The central problem with many large pharmaceutical companies' approach to off-label medicine is that they are marketing products that lack proper safety and efficacy assessments as though they had the same value as those that do have such data. My view of pharmaceutical value is that it is all about the level of health benefit, not the cost of making a product. As such, off-label medicine cannot be defended. However, in mitigation of the practice, some of the examples (though by no means all) were eventually developed further and properly approved for what had previously been off-label uses. Gabapentin, for example, was ultimately approved for the treatment of neuropathic pain caused by shingles – but not for any other off-label uses for which it was illegally promoted. I am perfectly willing to accept that there are examples where the off-label medicine is safe and effective. It is just that, absent the proof, it is very difficult to know which medical hypotheses will work out and which will fail. As we shall now see, there are some assumptions about drugs working in closely related areas to those in which they were approved, which turn out to be quite surprisingly false.

CHAPTER 5

Do no harm: Safety and efficacy

The essential message of this chapter is that off-label medicines involve a higher risk of adverse drug reactions (ADRs) than properly licensed alternatives and are less likely to be efficacious. How does this arise? A doctor's perception about the validity of prescribing off-label is often informed by his or her concept of relatedness. It is this principle which runs through this chapter, and I will point out reasons why the preconception is not (or may not be) valid. A practitioner's hypothesis is not necessarily more valid than any other therapeutic hypothesis from a medical researcher or a pharmaceutical scientist; these people also tend to believe that their theories are correct and are even able to convince other people of the merit in investing time and money in pursuing their ideas. Pharmaceutical companies often embark on drug development campaigns to uncover secondary uses, there being sufficient consensus and commitment to justify the expenditure of very large amounts of money over long periods of time: alas, for the vast majority of these programmes, when put to the test, 'the best laid schemes o' mice an' men gang aft agley' (as in Robert Burns' poem 'To a Mouse'), and the developments are abandoned before a product can be commercialised. The failure rates are not as high as de novo R&D, but they are nevertheless significant: historical analysis suggests that 25% of developments around secondary uses for existing compounds are successful, compared to the 10% statistic applicable to new drugs [100].

As far as safety is concerned, a variety of studies have shown an increase in risk associated with off-label prescriptions as opposed to licensed alternatives (see Table 5.1). As far as the overall relative risk is concerned, a series of studies have identified greater proportions of ADRs in paediatric patients prescribed off-label medicine, as opposed to products used in accordance with the approved status. The relative risk varies from 1.53 to 3.44, with the average roughly equating to slightly more than a doubling of risk for such patients. It is important to recognise here that we are talking about all kinds of adverse effects here, from the very mild to the very severe. Theoretically, this observation could be a result of an increase in very minor side effects, in which case the import of such observations might (rightfully) be discarded as 'unimportant'.

However, when we look at the types of ADR involved, again it seems that more of the ADRs associated with off-label medicine are serious compared to approved drug treatments. A study based on an analysis of the Danish national ADR database for children and adolescents from 1998 to 2007 showed that 60% of ADRs associated with off-label medicine were serious compared with a figure of 35% for drugs administered in accordance with their label [20]. A study from 2004 of ADRs in the United Kingdom drawn from paediatric patients reported that 35% of recorded ADRs involved off-label or unlicensed medicine and that 6 out of 10 medicines implicated in fatalities were off-label or unlicensed [117]. These reports have been picked up by

Off-label Prescribing – Justifying Unapproved Medicine, First Edition. David Cavalla.
© 2015 John Wiley & Sons, Ltd. Published 2015 by John Wiley & Sons, Ltd.

Table 5.1 Summary of studies showing increased risk of ADRs in paediatric patients treated with off-label medicine compared to treatment in accordance with approved status.

Study description	Country	Study year	Relative ADRs*	References
1388 paediatric inpatients	United Kingdom	2013	2.25	[113]
272 paediatric inpatients	Brazil	2008	2.44	[116]
1419 paediatric outpatients	France	2002	3.44	[16]
936 paediatric inpatients	United Kingdom	1999	1.53	[18]

*Adverse drug events (ADEs) describe any untoward medical occurrence that happens during drug treatment whether or not caused by the drug. ADRs form a part of ADEs, being defined as an unintended noxious response to a drug.

the EMA, who pointed out a need for better scientific evaluation of the use of unlicensed or off-label medicines in children and further analysis of the specificity of paediatric ADRs. They also noted that the real situation may be worse, since they were concerned about underreporting of paediatric ADRs associated with the use of unlicensed or off-label medicines [118].

The association of off-label use with excess ADRs is not confined to paediatric medicine. A study from a French regional drug monitoring centre among the general population (both children and adults) reported that more off-label prescriptions resulted in an ADR compared with approved medicines [119]. In a later study, 45% of patients with an ADR had been given an 'inappropriate' prescription. Although the control rate of off-label prescriptions is not reported here, we have seen a consensus level of one in five prescriptions from multiple other authors [120]; taking this figure, the risk of ADRs associated with off-label prescriptions would appear to be more than double that of approved medication [121]. Finally, we see a similar situation in medical devices, where the off-label use of sirolimus stents (to keep open a previously blocked blood vessel) is associated with a threefold higher rate of adverse cardiac events than on-label use [122] and similarly for biliary stents, which help to keep open the bile duct if obstructed. These devices are often used off-label in the peripheral vasculature, such that approximately 1 million off-label uses of biliary stents occurred from 2003 through 2006 and are associated with a truly alarming 81.2% of the adverse events [123].

Overall, a review from 2012 concluded rather cautiously that 'the results of previous studies have indicated that there may be some association between off-label and unlicensed medicine use and ADR risk', although it also called for further clarification on the situation [124]. Despite this cautious conclusion, the concordant previous evidence cited here calls for far more concern about the general use of off-label medicine. Not only are adverse events more commonly associated with the practice, but the events that are seen are of a more serious nature. The present level of evidence is generally to prove an association, rather than prove a causal relationship. It could be that off-label medicine is used in more serious conditions, where there is no licensed alternative. More work is needed to clarify this possibility, but at present, there is more than enough evidence to be quite concerned.

The issue of safety cannot be taken in isolation. In addition to safety, we shall be talking later in this chapter also about the efficacy of off-label medicine, which is also a concern. In most cases, there is a paucity of evidence that off-label medicines actually work. Some of the individual examples mentioned earlier pose issue of both safety and efficacy, like that of NovoSeven™, which was associated with both ineffective and unsafe treatment in an off-label setting. As a gross generalisation, one can say the following three things:

a When the off-label use involves unapproved patient groups, such as paediatric medicine, the main issue is likely to arise from the difference between children and adults, and often, this depends on dose, which has a likely impact on safety.

b When the off-label medicine involves solely a change of therapeutic use, the issue is likely to be lack of efficacy.

c When the off-label medicine involves a change of dose or formulation, safety and efficacy may both be affected.

This is a gross simplification because quite often (for instance, in the use of antipsychotics for behavioural problems in children) the medicine is off-label from the perspective of both patient population and indication. Where the safety from this dual off-label situation has been analysed, the relative risk of an ADR is 4.42 compared to the risk from an on-label alternative, suggesting this to be a clearly unsafe practice [16]. Moreover, as mentioned earlier, there is also an association between high levels of ADRs and off-label use among the general population, of both adults and children. It is also a simplification because the dosing in children may give rise to either an overdose or an underdose. Whereas an overdose is likely to give rise to increased risk of toxicity, underdose is likely to give rise to an increased risk of inefficacy. And finally, it is an oversimplification because using a drug that we consider safe for one use does not mean we consider it safe for all uses.

With these general principles in mind, let's now look in detail at why these problems arise.

Relative safety

Different therapeutic uses

It is commonly but mistakenly believed that once a drug is approved for one use, it is therefore safe for all uses. For instance, in proposing a widening of the reimbursement rules for Medicare, the organisation Medicare Rights Center made the argument that Actiq™ (fentanyl), which was indicated to treat cancer-related pain, should also be made available for the treatment of non-cancer-related pain states. They maintained that it was '...nonsensical for an opioid analgesic to carry an indication specifically for cancer pain. ...Actiq...[is being used]..."off-label" in well over 90% of prescriptions written. Why? Because it works' [125].

The Medicare Rights Center is not the only organisation to have used arguments like this to support wider use of drugs beyond their regulatory approval. But it is not just a question of 'does the drug work', since some of the possible ways in which an off-label prescription could differ from the approved label can compromise safety. In the case of Actiq™, the restriction of its use to cancer-related pain is a way of focussing the use of a strong, opiate-based drug on patients in severe pain who are much less likely to abuse it than if it were generally more available. So, even if Actiq is effective for the treatment of a painful bruise, there are more appropriate and safer agents. It is also the case that opiates do not work terribly well in certain types of pain, such as neuropathic pain or pain due to nerve injury.

It turns out from the experience with oxycodone, another strong opiate that has become much more widely used beyond just the treatment of cancer pain, that abuse has become a major problem. Deaths from overdoses of prescription painkillers are running at almost epidemic levels in the United States (about twice as many as pedestrians killed in road traffic deaths) [126]. Restricting the use of such agents to patient populations where the efficacy is assured and the side effect profile justified, or where there is no other option, is therefore important for patient safety.

Safety in one indication or patient group does not indicate adequate safety in all, and the balance of safety and efficacy needs to be borne in mind before an off-label prescription can be validly written. Severe conditions can tolerate more severe side effects than mild ones. So, cancer patients treated with chemotherapy suffer a whole range of severe side effects that would be considered unacceptable in a milder or less serious condition. This type of issue comes up time and again in off-label uses. Here are some examples of off-label uses of drugs that require a recalibrated safety-efficacy assessment in their new indication, as a result of the relatively less serious nature of the secondary indication compared to the primary utility for which the drugs were originally licensed: antidepressants for hot flashes, the anti-narcoleptic drug modafinil for general tiredness, β-blockers used regularly for stage fright, buprenorphine (an opiate) for depression or the anti-malarial drug quinine in treating night-time leg cramps (see p. 88).

So, in the regulatory approval process, different standards are applied in different therapeutic indications. How do we know this, rather than just suspecting it? Well, in 2014, FDA Commissioner Margaret Hamburg said: 'People with serious or life-threatening illnesses, particularly those who lack good alternatives, have told us repeatedly that they are willing to make some trade-offs in order to gain access'. As a consequence, the FDA will look favourably on smaller trials where the condition is serious and safety less of a priority. She continued, 'There is no reason to expect drugs to be tested on similar numbers of patients [in order for approval to be granted], regardless of the disease' [127].

This is one of the particular problems with the off-label use of antipsychotics: these drugs are approved for serious mental illness like schizophrenia. Their use in childhood behavioural problems is not justified, even if efficacy were determined to be good, without a re-evaluation of the safety and tolerability of such agents in different patient groups. A similar issue pertains to the use of the same agents in older patients, where they have been widely used for behavioural problems in dementia victims, as a 'chemical cosh', or for insomnia, another popular off-label use of antipsychotics.

Gwen Olsen, the campaigner against pharmaceutical marketing referred to earlier (p. 63), gives an anecdotal story about a spirited old lady she saw (and rather admired for her energy) during her visits as a pharmaceutical rep to nursing homes in Texas. One day she noticed the old lady was absent and asked what was the matter with her. She was given a vague non-committal answer by the staff at the home but later saw the lady in question, slumped in a wheelchair, with drool emanating from her mouth, unable to recognise or communicate with anyone. She had been administered an antipsychotic. Though an individual anecdote, it is clear that the extent of this and similar practices is widespread. As pointed out in Chapter 1, off-label prescription runs at about 60% in psychiatric medicine, and the atypical antipsychotics which are used for this type of medicine are among the most widely prescribed of all drugs, with aripiprazole (Abilify™) heading the list of top pharmaceuticals by value in the USA for 2013. It would be logical to conclude that many of these antipsychotic prescriptions are for non-approved uses or patient groups.

Sometimes additional risks are associated with particular characteristics of the patients in a new therapeutic area. For example, as discussed in Chapter 2, it is true that the use of tricyclic antidepressants for the treatment of neuropathic pain has been known for years – their efficacy is well established. However, that level of evidence is still less than acquired during the formal regulatory trials in support of approved status. Such trials will establish on a large scale the safety of these agents in a new patient group. In the particular case of neuropathic pain, this condition is often associated with diabetes, which carries a particular cardiovascular risk. Treatment of painful diabetic neuropathy with tricyclic antidepressants would need to take into account the cardiovascular side effects of these drugs, which have been linked to increased blood pressure, weight gain and diabetes [128]. The use of a drug which carries a certain elevated cardiovascular risk in a patient group at heightened risk of cardiovascular disease is a concern that would almost certainly be raised by a regulator, yet has not been raised in consideration of the clinical trial evidence base so far available.

A similar issue applies even when we consider antibiotics, which are generally considered to have a good side effect profile. Here again, their acceptance depends on the seriousness and degree of utility against the disease they are being used to treat. When a drug is not efficacious or only efficacious to a very limited extent, the risk–benefit profile becomes extraordinarily sensitive to safety. And since we know there is always some risk associated with a prescription, the efficacy cannot be zero for the medicine to be justified [129].

So why are antibiotics used for viral infections such as the common cold? Airway respiratory infections (ARIs) are the primary reason for antibiotic prescribing in the United States and account for about 50% of all antibiotic prescriptions to adults [33]. Antibiotics are not labelled for the treatment of viral disease, yet many doctors prescribe them in this condition despite the lack of medical evidence for their effectiveness. It is often assumed that the reason for this is the difficulty of correctly diagnosing the cause of the ARI, yet when subjected to an assessment, doctors are actually very accurate in their ability to correctly identify different kinds of ARI using their clinical judgement, with high rates of sensitivity and specificity [130].

Another mistaken belief is that antibiotics prevent secondary bacterial infections when the primary cause is viral. Though there is some evidence for their utility in measles, their use in ARIs is not advised. The well-respected Cochrane group concluded 'there is not enough evidence of important benefits from the treatment of upper respiratory tract infections with antibiotics and there is a significant increase in adverse effects associated with antibiotic use' [131]. A similar conclusion was reached by the Cochrane group in regard to the treatment of bronchitis with antibiotics, namely, that there was evidence limited to a modest benefit in some subsets of patients, but that overall this benefit should be seen in the context of a self-limiting condition, the involvement of some drug-related side effects, increased resistance to pathogens and significant costs of medication [132]. There is some evidence in the United Kingdom that there are regional variations in antibiotic prescribing, reflecting differences in the attitudes of doctors rather than a variation in pathogenic exposure. Rates of prescriptions in the North East of the United Kingdom were significantly higher than the South East, a variation that was termed 'inappropriate' by the NHS chief pharmaceutical officer [133]. In summary, a significant proportion of antibiotic prescribing for ARIs is therefore inappropriate, exposing individual patients to the risk of adverse drug events and increasing the prevalence of antibiotic-resistant bacteria. Also, it increases cost, with unnecessary antibiotics accounting for over $1.1 billion of the amount spent annually on ARI treatment [134].

Although it is well known that certain patients are allergic to beta lactams, more serious side effects can occur with other antibiotics such as the aminoglycosides.

Examples include gentamicin and neomycin, whose use is limited by the risks of ototoxicity and nephrotoxicity. These risks have been well known for many years. More recently, the risks associated with a newer class of antibiotic, the fluoroquinolones, have become better understood. The best known are Cipro (ciprofloxacin), Levaquin™ (levofloxacin) and Avelox (moxifloxacin). In 2010, Levaquin was the best-selling antibiotic in the United States.

Unfortunately, instead of being reserved for use against serious, perhaps life-threatening bacterial infections like hospital-acquired pneumonia, fluoroquinolones are often inappropriately prescribed for sinusitis, bronchitis, earaches and other ailments that can be treated with safer but less potent alternatives, or may even resolve without intervention. They are sometimes even used for viral infections, which are not susceptible at all to antibiotics.

Fluoroquinolones carry a 'black box' warning mandated by the FDA that informs prescribers of the link to tendonitis, tendon rupture and muscle weakness; in 2013, neuropathy was added to the list. A further study published in 2012 in the *Journal of the American Medical Association* documented a 4.5-fold higher risk of incurring retinal detachment (potentially leading to blindness) in patients treated with fluoroquinolones, compared with non-users (although that report has been challenged) [135]. In another study, a significantly increased risk of acute kidney failure among users of these drugs was recorded [136]. In addition to these serious side effects, there are idiosyncratic toxicities associated with fluoroquinolones.

Guidelines by the American Thoracic Society state that fluoroquinolones should not be used as a first-line treatment for community-acquired pneumonia; the Society recommends that doxycycline or a macrolide antibiotic be tried first. However, in 2003, a 23-year-old nursing student was prescribed the fluoroquinolone moxifloxacin for a persistent cold that was diagnosed as acute bronchitis (not an approved indication), as a result of which the young woman developed a severe adverse reaction that led to liver failure and death in a few days. In 2012, the *New York Times* ran the story of a 33-year-old Manhattan resident and website manager for City College of New York, Lloyd Balch [137]. He was treated for a mild case of pneumonia with levofloxacin, following which he developed some serious adverse musculoskeletal, neurological and vision problems.

These individual anecdotes are meant to colour the possible outcomes, rather than to indicate a limitation on their frequency. In fact, there are plenty of accusations related to the adverse event profile of these drugs: in 2011, fluoroquinolones were the subject of more than 2000 lawsuits from patients who had suffered severe reactions after taking them. The increased use of fluoroquinolones has also been blamed for increases in antibiotic-resistant *Staphylococcus aureus* (known as MRSA) and severe diarrhoea caused by *Clostridium difficile*. One study found that fluoroquinolones were responsible for 55% of *C. difficile*-associated diarrhoea at one hospital in Quebec [138].

In summary, the central message is that a patient with a dangerous disease may be justifiably treated with a dangerous drug; a patient with a non-serious condition demands a much safer regimen for their treatment. But generalities aside, sometimes the problems can be very difficult to predict, and you only really know if a drug that is safe in one indication can be used in another if you test it.

Tiagabine is an anti-epilepsy agent indicated as an adjunct treatment for partial seizures in adults and children over 12, but it was also widely prescribed off-label for the treatment of neuropathic pain, anxiety or bipolar disorder; during the 6 years to 2004, the level of such off-label use rose from 20% to 94%, and the annual number of dispensed prescriptions rose 20-fold, to about 1 million. However, when it was used off-label in non-epileptic patients, it was associated with the *development* of seizures, in some cases just days after initiating the drug [139]. The reason is complex: when used as an adjunctive epilepsy treatment, patients were by definition also taking another anti-epileptic, and in consequence their levels of drug-metabolising enzyme (cytochrome p450 3A4) were often raised – this is because the primary anti-epileptic induces the metabolising enzymes. (In effect, the patient's body is recognising a foreign chemical and seeking ways to eliminate it more effectively; we see the same thing with alcohol: the regular drinker has a higher tolerance due to higher levels of metabolising enzymes.) Tiagabine is also metabolised by this enzyme, so blood levels in epileptics would generally be less than in non-epileptics. Off-label use therefore generally involved exposing neuropathic pain patients, for example, to higher doses of tiagabine than were safe, resulting in uncovering the seizure-promoting effect of the drug.

Safety of a drug is associated with the use to which it is put and cannot be assured more widely without testing. Despite the insufficient general awareness of this issue, there is one group who are very attuned: the regulatory authorities. An excellent example of this relates to the story of alemtuzumab, a biological product that was first given the trade name Campath-1H. This originated from the pathology laboratory of the University of Cambridge (hence the name) and was further refined at the Laboratory of Molecular Biology, also in Cambridge (the same place where monoclonal antibodies were first discovered by Milstein and Kohler). It was first approved for the treatment of chronic lymphocytic leukaemia (CLL), both in Europe and in the United States, and until recently remained a successful and uncontroversial treatment for this condition. It works by binding to the CD52 receptor, which is present on the surface of mature lymphocytes but not the stem cells from which they are produced. Binding to this receptor causes the cells to embark on a process of programmed death (apoptosis); in a disease such as CLL, which involves uncontrolled expansion of leukocyte populations, this is obviously a valid therapeutic approach.

Shortly after it was first discovered, its effects in multiple sclerosis (MS) were identified. MS is a disease in which the body develops an immune reaction against the patient's own nerve cells, orchestrating an attack on the myelin which acts rather like the plastic insulating sheath on an electric cable. In MS, alemtuzumab depletes the body of lymphocytes, which causes a reprogramming of the immune system: in other words, the body's lymphocytes are mostly removed, but the stem cells from which they derive are left intact (since they do not express the CD52 receptor). The body's immune system undergoes what is essentially a process of 'rebooting', producing a new, modified repertoire of immune cells which no longer regards myelin and nerves as foreign, thus quelling the autoimmune pathophysiology of MS. This can result in the prevention of a relapse in disease for MS patients lasting from 3 to 5 years – a really dramatic result. In addition, many patients experience a reduction in disability which is far beyond that seen for standard (interferon) treatment.

The complexities of optimising the correct timing and dosing for alemtuzumab in MS resulted in a 20-year clinical development period stretching from the early 1990s, when this use was first proposed by Alastair Compston, to 2011 when the results from the second of the two regulatory trials were announced. The optimised timing refers both to initiating the treatment in early rather than late disease, and to the administration of two doses of the drug, separated by a period of between a year and 18 months. Sanofi, the manufacturers of the product, decided to withdraw the original cancer product and focus entirely on the new use of alemtuzumab for MS (with a new trade name, Lemtrada™). The reasons for this were mainly commercial, because the pricing for MS treatments is higher than the first leukaemia product cost, and this ploy by the manufacturers drew substantial criticism from various industry watchers. (On the other hand, the development of Lemtrada took a very long time and cost over $1 billion in actual development expenditure.)

In the course of the MS trials, a strange new toxicity was observed which had not been apparent in leukaemia patients previously treated with the product. In about 25–30% of patients, another serious autoimmune disease developed, most frequently involving overactive thyroid function. Known alternatively as Graves' disease, it is brief and treatable, but nevertheless it is serious. Its occurrence was entirely unexpected (unless you can say that after reprogramming the immune system, nothing is unexpected). On the other hand, in terms of efficacy, Lemtrada was spectacularly successful, offering disability improvement at least to 5 years after treatment, and significantly more effective in suppressing relapses than interferon beta-1a. The company were successful in persuading the European regulatory authorities that the product could be approved in September 2013, in other words that its serious risks were worth its serious benefits in a serious disease. However, later that year, the FDA took the opposite decision and denied approval, at least for now. Though this decision has been criticised by various MS patient groups, it emphasises the fine risk–benefit decision that regulators must sometimes make. Part of the reason for the denial relate to the unanticipated risks of *using Lemtrada in the MS population*: risks that did not apply to using Campath-1H in the leukaemia population.

In addition to this surprising patient- and disease-specific toxicity and without taking anything away from the perceptiveness of the medical profession, it is impossible to expect the same degree of analysis in a general practitioner (GP), physician or even a hospital doctor, as that allocated by the regulatory agencies. Jerry Avorn, who I referred to earlier, said 'It is unrealistic to expect each physician to have the time and expertise to subject such claims to the same kind of scrutiny that the FDA exercises when it reviews a drug application or a request for a new indication. The complexity of the assessment that is required, along with the high stakes of getting the assessment wrong, provided the rationale for having a formal drug-approval process in the first place'.

We should also remember that a typical GP consultation in the United Kingdom requires the doctor to diagnose and reach a prescribing decision in 7 minutes. The regulatory bodies certainly have their faults, and they do not always get it right, but overall, the public have every reason to regard their judgement as the gold standard – certainly more reliable and objective than a pharmaceutical marketing campaign to enlarge treatment options for the latest, expensive drug. This is why I think we should continue to place off-label medicine into the 'last resort' category – to be the exception rather than the rule – and find means to limit the extent to which it is applied on an unsuspecting public. At the same time, instances where off-label medicine has been used successfully should be properly evaluated so that they can be applied with confidence for the greatest benefit to the greatest number of patients.

Chronic versus acute dosing

There are plenty of examples where drugs which are considered safe for acute (short-term) usage can produce safety concerns when they are given for extended periods. For example, Duract™ (bromfenac) was an analgesic, labelled only for treating acute pain and only for short-term use of less than 10 days [140]. However, some physicians prescribed Duract off-label for longer durations, despite efforts by regulators and the manufacturer to educate physicians about the dangers of doing so. Unfortunately, when used in this way, it caused liver failure and was withdrawn from the market less than a year after approval.

A similar situation evolved in the use of chronic 5-HT3 antagonists for irritable bowel syndrome (IBS). Drugs in this class have been used for decades in the prevention of cancer chemotherapy-induced nausea and vomiting. In this utility, agents such as ondansetron are safe and effective, generally given acutely around the administration of the chemotherapeutic agent. However, GSK noticed that 5-HT3 antagonists could also be useful in the treatment of IBS, particularly of the diarrhoeal type. This condition can cause substantial morbidity and a high amount of healthcare allocation; but although unpleasant, IBS is not a life-threatening condition. It is also a substantially heterogeneous condition, with both constipative and diarrhoeal as well as mixed types, with substantial differences in severity and cause. GSK developed a new drug, alosetron, and showed it to be clinically effective in female patients with diarrhoea-predominant IBS in large randomised controlled clinical trials (for some reason, it was not found to be useful in males). However, after approval in 2000, post-market surveillance showed that several serious adverse effects related to severe constipation or ischemic colitis, including death, occurred in some unfortunate alosetron-treated patients. IBS requires chronic rather than acute treatment, and this safety issue led to alosetron's withdrawal from the market; it was subsequently reintroduced but only accompanied by a careful clinical surveillance programme and severe regulatory restrictions [141].

A more serious situation, which ultimately led to a scandal, was found with the appetite suppressant Pondimin (fenfluramine). Fenfluramine works via serotonergic pathways to induce a feeling of satiety. Again, this was also approved only for short-term appetite suppression; however, it was widely prescribed with another appetite suppressant, phentermine, and used long term for the treatment of obesity. Phentermine had the effect of opposing the fatigue associated with fenfluramine, which as a sole agent had not been very successful because its moderate results were overshadowed by uncomfortable side effects, including drowsiness and altered moods. Phentermine could combat these side effects. Even though the FDA had not approved their combined long-term use in this way, in 1996, the total number of prescriptions for fenfluramine and phentermine ('Fen–Phen') in the United States exceeded 18 million.

In 1997, Dr Heidi Connelly at the Mayo Clinic reported 24 cases of unusual heart valvular defects in patients taking the Fen–Phen combination [142]. They had been doing so for periods as long as 28 months. As a result of these findings and another 75 cases of heart-valve disease reported to the FDA, all fenfluramine-containing products were withdrawn from the market in September 1997. The FDA concluded from its own data and from Dr Connelly's study that 30% of the evaluated patients had unusual echocardiogram results.

This was not all. In addition to the cardiovascular problems, the d-isomer of fenfluramine, dexfenfluramine, also increased the risk of primary pulmonary hypertension, particularly when patients received high doses for more than 3 months. In these people, the risk of contracting the condition is increased 23-fold, from 1 in 500 000 to 1 in 20 000. Primary pulmonary hypertension is a serious condition that tends to affect women in their 30s. After diagnosis, patients with this disease typically die within 2 years.

Litigation related to Fen–Phen has been widespread. This is partly because of the numbers of prescriptions – there are estimates that between 6 and 7 million people took it in the United States – and also because of claims related to the latency of the cardiovascular or pulmonary effects years after cessation of the treatment. Pfizer, who took over the Wyeth unit that manufactured Fen–Phen, set aside $21 billion to settle a tsunami of litigation that at one point numbered over 175 000 claims and was still involved in legal action 15 years after the drug was withdrawn.

The regulatory agencies are aware of the different risk profiles of chronic medication relative to acute. In 2005, the biotech company Vernalis was developing frovatriptan for the treatment of migraine. It was similar to analogous products like sumatriptan, eletriptan, almotriptan, zolmitriptan and rizatriptan (together, these are the 'triptans'), all of which activate a receptor that constricts blood vessels in the brain and relieves swelling, thereby relieving migrainous pain. Thus, there are half a dozen triptans used for migraine, reflecting a widespread degree of satisfaction with the safety and utility of this class of drug in this condition. As a class, they have been available for over 30 years and are considered sufficiently safe that in the United Kingdom one of them can even be obtained over the counter, without a prescription.

Frovatriptan was longer acting than the others, however, and the people at Vernalis wondered if it could be used prophylactically to prevent the regular migraines associated with the menstrual cycle. Menstrual migraine occurs in 40–60% of female migraineurs around the time of menstruation and is therefore quite a debilitating monthly problem. Studies showed that in women with chronic migraine, the frequency and duration of migraine attacks, and the associated disability, were greater in the 6 days after the onset of menses. However, the exact time of onset within this period is unpredictable, so a longer-acting agent has distinct advantages compared to a shorter-acting one. Premenstrual migraine was given a particular definition, as 'a migraine without aura that occurs in at least two-thirds of menstrual cycles during the 5-day perimenstrual period'.

Vernalis had already developed frovatriptan for migraine *per se*. However, they saw a significant additional commercial opportunity if they were able to gain approval for the prophylaxis of menstrual migraine. All of the triptans have been studied, and shown superior to placebo, in the acute treatment of menstrual migraine. However, the prophylaxis of menstrual migraine involves a subchronic form of administration, often begun 3–5 days before the onset of menses and continued for 3–7 days into menses.

Vernalis studied frovatriptan in a large trial involving 546 women, administering it prophylactically in a couple of different dosing regimens over the menstrual period; in this trial, it was shown to be effective in reducing the frequency of menstrual migraines compared to placebo treatment. Buoyed by this success, the company went on to conduct further studies and, in July 2006, filed a regulatory application to the FDA supported by data from four trials. These included two Phase III studies examining the efficacy and safety of once- and twice-daily dose regimens of frovatriptan in the short-term prevention of menstrual migraine; a pharmacokinetics and tolerability study of once- and twice-daily dosing of frovatriptan; and a 12-month, open-label safety study evaluating a 6-day dosing regimen of frovatriptan in 525 women [143].

It took the FDA over a year to assess the application and to respond. When they did, in September 2007, they rejected it as 'non-approvable' and cited safety as a significant component for their decision. Specifically, they said that even though serious vascular adverse events were not observed in this drug development programme, an increased risk (compared to the approved acute use) could not be ruled out. They also raised concerns about the clinical meaningfulness of the effects observed in Vernalis' studies. The FDA rejection does not prevent the off-label prescription of frovatriptan for the prophylaxis of menstrual migraine, or indeed any other triptan, but it does indicate the dissatisfaction by the regulatory agency of the chronic use of a medication on safety grounds, even though its acute use is considered acceptable. Indeed, safety in acute use is considered satisfactory for all of the triptans.

At the time of this event, frovatriptan had been on the market for under 6 years. However, regulators do not just add warnings on safety regarding chronic use for new drugs; sometimes they add them for old drugs too. In 2009, they reviewed the chronic use of metoclopramide, a drug that had been used as a treatment for various stomach and digestive disorders since 1964. As a result of the review, they then released a warning. Whereas it was 'approved for the short-term (no longer than 3 months) treatment of gastrointestinal disorders', an analysis of prescription claims data had shown that around 15% of patients received prescriptions for metoclopramide for more than that [144]. The FDA warned that 'frequent and long-term use of metoclopramide has been linked to tardive dyskinesia', a debilitating movement disorder involving involuntary, repetitive movements of the extremities, grimacing or tongue protrusion. They further warned that symptoms of tardive dyskinesia may be permanent and the chances of its reversal diminished as the dosage or length of time the medication had been taken increased.

All in all, it is clear that drugs which are considered safe for acute use are not necessarily safe for chronic use. Off-label prescription which involves the use of drugs for longer than their label indicates should be considered inadequately assessed for safety.

Different dose

It is rather obvious that a lower or higher dose of a prescription medicine than is on the label could produce safety issues. This issue becomes of particular relevance if different uses or patient groups involve different doses or dosage regimens. There are a couple of different scenarios here: simply, it may involve administering a drug at a higher dose than approved; another possibility could involve administering a drug that is labelled for three times a day just once a day, but at three times the individual dose. Indeed, we saw with the erythropoietin-stimulating agents an increased level of thrombogenesis in patients treated with higher doses of these agents (see p. 58) [145].

We may see this issue crop up in future assessments of anti-diabetic drugs like liraglutide (Victoza™), one of a class of GLP-1 agonists, for the treatment of obesity. Diabetes and obesity often co-occur, but the clinical effect of liraglutide in the treatment of obesity was only modest at the currently approved doses. The greatest weight loss effect required doses up to 2.5 times the normal dose of 1.2 mg daily [146]. Although the drug has been proposed for the treatment of obesity, it is currently unapproved for this purpose; while some of the dose-related side effects are mild or moderate, like vomiting and diarrhoea, other side effects that have been noted, albeit rarely, include pancreatitis, which is of much greater concern. The potential therefore exists for overweight diabetics to use the drug in high doses for its weight-reducing properties, while running the risk of more serious dose-related adverse events.

Let's focus on a couple of further examples. Factor VIIa (NovoSeven) is an expensive and potent procoagulant, approved by the FDA in 1999 for intravenous use in haemophiliacs. The approved dosages can vary up to 120 µg/kg. However, I mentioned earlier that it has been used extensively off-label, with reports that the incidence of this practice has been as extreme as 97% (see p. 14). Some of the off-label use exceeds the approved dosage. This is important since it would be expected for a procoagulant to cause clots to form in proportion to the dose, and excess doses could cause ADRs like stroke, deep vein thrombosis and heart attacks. Despite this obvious safety risk, the product has been used in doses of up to 400 µg/kg, namely, between three and four times the maximum approved dose. The use of off-label doses higher than those approved on the label has been correlated with a doubling of the risk of arterial thrombosis. Approximately 1 in 16 such patients suffered an arterial thrombosis, compared to half that level in patients dosed within the approved range [48]. The costs of factor VIIa at the dosage of 90 µg/kg body weight every 2–3 hours may exceed $50 000 per day, whereas at 400 µg/kg the costs are over $200 000 per day. That's a pretty expensive way to die.

Still in this field, enoxaparin, a low-molecular-weight heparin, is an anticoagulant used in coronary care settings. It is indicated for the treatment of acute coronary syndromes, otherwise known as unstable angina and characterised by episodes of chest pain at rest or with minimal exertion. As an anticoagulant, not only is there a greater risk of clotting leading to a stroke or heart attack associated with a high dose, but if too little is given, then it would have a sub-therapeutic effect. It is therefore surprising that in a 2007 study almost half of patients who received enoxaparin were given doses that were either higher or lower than the labelling recommends. Though this may more reasonably be ascribed to chaos than conspiracy, we are more interested in the results of this pattern of prescribing than the reasons why it occurred. An analysis of outcomes in hospital patients treated with the drug showed an excess dose was significantly associated with a 43% greater risk of major bleeding and 35% greater risk of death compared with a recommended dose [147]. Enoxaparin is a generic drug in the United States of which 15 million units were prescribed in 2012, making it the 15th most frequently prescribed product.

Getting the dose (and dosing interval) of enoxaparin right requires a little care and attention, since patients with poorly functioning kidneys need less of the drug, as they excrete it more slowly. A blood test should be used to measure kidney function, and body weight, height and sex also come into the calculation. Nevertheless, this is all you need to calculate the right dose of the drug; there really isn't a good excuse for getting it wrong when a patient's life is at stake.

Things are a little more complicated when it comes to children.

Differences between children and adults

Most approved drugs allowed on the market have established a favourable balance between beneficial and harmful effects for adults only. In 2003, approximately 75% of the drugs licensed since the early 1970s had been approved without paediatric drug labelling. That proportion stimulated the development of regulatory incentives for industry to prioritise paediatric studies and approvals and has resulted in a significant improvement in this situation, with the 500th paediatric label change made in 2013 [148]. It is now the norm rather than the exception for drugs to be approved for children, but these changes affected new medicines, rather than old. Paediatric medicine therefore still often leaves doctors with little alternative than to deploy an off-label prescription, even though there may well be insufficient information about safety and efficacy in this particular group; this is particularly so in neonates. This difficulty is brought into focus by the fact that around 60% of important antidotes used in the treatment of poisoning, which may very well be an emergency situation, have to be prescribed off-label in children. In addition to lack of information, specific paediatric formulations are often not available.

As discussed earlier, studies in various paediatric hospital environments have shown a considerable proportion of medicines used outside the product licence. But in addition, there were some examples where the drug comprising the active ingredient had not received any licence at all (an unlicensed drug). A study based on over 66 000 dispensing records for the year 2000 in Dutch children showed that unlicensed drug use is highest among 0- to 1-year-olds (at 35%), whereas off-label medicine is highest in 12- to 16-year-olds (at 27%) [149]. Taken together, among all age groups of children, the proportion of off-label and unlicensed drug uses was 37% of all prescriptions. However, beneath the overall figures, certain areas showed even higher levels of unlicensed or off-label use. For instance, the proportion of unlicensed diuretics was over 70%, and the level of unlicensed steroid use was nearly 60%. As for off-label use, this was particularly high (over 80%) among anti-migraine, hypnotic and sedative drugs and nearly 75% for eye and ear drugs. Given that these statistics involve various drugs with some serious side effects, it is of huge importance for patient safety that the doses are correctly calculated. But the problem is that in paediatrics, the patient may be a premature infant weighing 500 g or a fully grown adolescent weighing as much as 100 kg. In addition to concerns about safety in overdose, we should also be concerned about lack of efficacy if the dose is too low.

Paediatric drugs have usually been used off-label by extrapolating efficacy, dosing, administration and side effect profiles from adult studies. However, children differ pharmacokinetically from adults, and medication for adults cannot simply be administered in smaller doses. There are plenty of examples from the history of paediatric pharmacology illustrating that newly marketed drugs may have incomplete adverse event profiles that make it inadvisable for widespread community utilisation in youths and children before uncommon or rare serious adverse events are known. Several cases illustrate the risks even for commonly used and accepted treatments.

In 1949, an important new antibiotic called chloramphenicol was introduced promising effective treatment of serious infections not controlled by other drugs that were then available. It could be given orally or by injection, and was easily synthesised, inexpensive and thought, at the time, to have no significant toxicity. However, in 1959, it was linked to the development of 'grey syndrome' in many of the infants so treated. In neonates (especially premature babies), inadequate levels of metabolising enzymes coupled with immature kidneys for excretion of the drug led to increased toxicity.

There is a surprising and substantial increase in chloramphenicol metabolism and excretion capacity during the first days and weeks of life, but in the newborn baby, toxic levels of chloramphenicol can be reached relatively easily. After less than a week, this can result in jaundice, vomiting, respiratory distress, ashen grey colour of the skin, limp body tone and cardiovascular collapse; ultimately, in many cases, 'grey syndrome' causes death. The conclusion (in hindsight) was that the pharmacokinetics of chloramphenicol should have been studied in neonates before it was widely used, since they proved to be substantially different even from other young children.

A similar issue involves the anaesthetic propofol, which incidentally was the drug involved in the death of the 'King of Pop' Michael Jackson in 2009. When used in paediatric intensive care units, this drug has been linked with 'propofol infusion syndrome'. This condition involves a potentially devastating cardiovascular and metabolic derangement characterised by the occurrence of lactic acidosis, acute kidney failure, breakdown of muscle tissue and circulatory collapse after several days of propofol infusion. It particularly affects critically ill young patients. Sildenafil too causes specific problems in children; it was mentioned previously for its use in the treatment of pulmonary hypertension (as well as erectile dysfunction). It is approved for adults in this lung indication, but not approved for children. And despite extensive clinical experience on its utility in this population, there is an apparent increase in mortality during long-term therapy, as a result of which the FDA has issued a warning against its use in this context [150].

Another example concerns promethazine, an antihistamine used for the treatment of allergies, contained in over-the-counter cough and cold products. However, in children less than 2 years of age, there have been reports of serious and potentially life-threatening respiratory depression associated with promethazine, resulting in a regulatory warning that recommends avoiding use in such patients. The warning appeared nearly 30 years after the product was first approved, demonstrating that issues of this kind can take many years to be identified.

Most physiological processes that govern the pharmacokinetics of drugs mature during the first year or two after birth. Some of these changes occur during the first days and weeks of life. Thus, we expect the youngest babies to be the most likely to experience the most aberrant responses. However, paediatric drug safety issues are not simply the consequence of immature enzyme systems in the neonate.

Elementary school-age children can be at increased risk of ADRs and can experience problems distinctly different from adults treated with the same drug. Phenobarbital was the first of the barbiturates to be introduced as an anti-epileptic more than 90 years ago. Currently, its long-term use in children and adolescents is rarely justified because it is now known to increase the risk in children of behavioural problems known as 'paradoxical hyperactivity'.

Another good illustration is tetracycline, a broad-spectrum antibiotic widely acclaimed and enthusiastically prescribed when it was introduced in 1955. However, it would take 8 years for a definitive paper to demonstrate that this antibiotic was responsible for causing thin and stained enamel of primary and secondary teeth. The dangers actually occur at any time when teeth are growing, from exposure in the womb during the last trimester of pregnancy up to 8 years of age. Tetracycline chelates (i.e. binds to) calcium, leading to loss of enamel. Before this safety issue was recognised, several million children were exposed to tetracycline even though alternative antibiotics would have been safer [141]. We also see a similar side effect associated with adolescents in a rapid phase of growth with depot-medroxyprogesterone acetate, which is sometimes prescribed off-label as a contraceptive, yet specifically reduces bone mass in this population yet to reach full growth [152]. These types of safety problems require centralised adverse event reporting: it is very unlikely an individual doctor would identify these risks based on his or her sole prescribing experience. As we discover in Chapter 8, this is an aspect of drug regulation called pharmacovigilance, upon which prescribing practice increasingly relies.

But perhaps the best (although controversial) representation of the problem of safety stratified by age involves suicidal behaviour and antidepressants. This issue was first identified in the proposed expansion of the use of paroxetine, a selective serotonin reuptake inhibitor (SSRI), for use in depressed children. At the time, in 1999, this GSK drug was approved for depression in adults. In the regulatory submission for paroxetine in the new population, the company submitted a report of a trial that included evidence of an increase in 'suicidal ideation/gestures' among 12- to 18-year-old adolescents treated with the drug. Rather than report this directly to the regulatory authorities at the time it happened (as they should have done), they bundled it together with the results from other trials in the later regulatory submission. Indeed, it is noteworthy that the trials in adolescents showed not just an increase in suicidal ideation but also a lack of improvement in the primary endpoint of depression. (It is also important to note that GSK misreported the results of this trial, publishing it in a major scientific journal with a long list of famous 'authors', most of which knew little or nothing about the trial, calling it a 'success' and burying the suicidal ideation problem under the category 'emotional lability'.) Later, GSK successfully managed to get regulatory approval for paroxetine in adults for other conditions such as post-traumatic stress disorder (PTSD), obsessive compulsive disorder, anxiety and panic disorder.

Without getting embroiled any further in the evil corporate conspiracy debate, I want to focus on the issue of suicidal behaviour and ideation, since this is a complex issue. First, it turns out the excess risk in this area with paroxetine is focussed on children, adolescents and young adults up to 24 years old; the rate among old adults treated with paroxetine is actually decreased relative to placebo. If we treat adults over 24, paroxetine is effective without an overall increased risk of suicidal behaviour. If we go off-piste and look to treating children, we have a non-effective antidepressant, with an increased risk of suicidal thoughts [153]. Second, suicide and thoughts of it are common in depression. A study from 2007 showed that suicide is attempted most often a month before treatment. Subsequent to the initiation of treatment of any form (whether pharmaceutical or psychotherapeutic), there is a marked decrease in such behaviour. This time course should be seen primarily as an *increase* in suicidal thoughts as a consequence of the hiatus or crisis in a person's life that often causes the depressed mood and need to seek out professional help. The graph shown in Figure 5.1 shows that, in under 25-year-olds, professional help reduces the likelihood of suicide attempt, rather than increasing it, whether that help is in the form of antidepressant drugs or psychotherapeutic counselling [154]. The study showed a similar result in older patients. This type of analysis shows the limitations of a bare statistic suggesting merely that 'paroxetine increases suicidality'. One may even compare this statement to the idea that one should stay away from hospitals, since so many people die in hospitals. And third, the absolute risk of suicide is different in depressed adults and children. Consider, for instance, the typical depressed menopausal mother of three, who will live in misery rather than orphan her kids, and compare to a headstrong teenage boy rent with anger and violence directed towards the outside world and inwardly in equal measure. Who is more likely to kill themselves?

These conclusions are bolstered by two further pieces of evidence I came across during the final stages of production of this book. The first is a study published in the *British Medical Journal* in 2014, reporting that attempted suicides by adolescents increased 21.7% 2 years after the FDA's "black box" warnings about suicidality and antidepressants, as the use of these drugs fell by 31%. Attempted suicides among

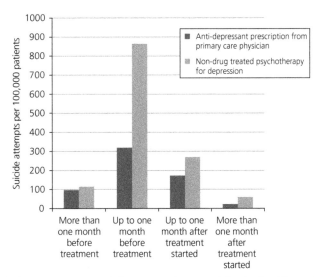

Figure 5.1 Suicide attempt among adolescents and young adults (under 25 years) before and after receiving new antidepressant prescriptions from primary care physicians or psychiatrists or starting psychotherapy. Drawn from Ref. [154].

people between the ages of 18 and 29 soared 33.7%. The researchers arrived at these statistics by analyzing insurance claims data from 11 companies, using reports of drug poisonings to determine suicide attempts; they conclude that safety warnings about antidepressants and widespread media coverage decreased antidepressant use, alongside a simultaneous increase in suicide attempts among young people [155]. The second is a seminal paper by a group of academic psychiatrists [156] who, for the first time, evaluated longitudinal data on individual subjects to look at suicide and thoughts thereof alongside depression, over their treatment period. The conclusions, based on two SSRIs, fluoxetine (Prozac) and venlafaxine (Effexor), are broadly that these drugs do diminish depression, do so more effectively in older adults than in younger ones, that suicide and thoughts thereof are correlated with depressive mood rather than SSRIs, and that suicide and thoughts thereof are far more common in younger adults than old ones.

So, depressed children and depressed adults behave quite differently, and many antidepressants (not just SSRIs, but also older drugs like amitriptyline, imipramine and desipramine) carry warnings about increased suicidality specifically in younger people [151]. These warnings, in fact regulatory warnings in general, are not always adhered to by the medical profession. A study of off-label paediatric use in Germany showed that 17% of the instances involved the prescriber ignoring recommendations on active ingredient, dose units or formulations for a specific age group. In this study, the authors noted that the prescription of quinolones to children was observed, even though there was a restriction regarding the use of quinolone antibiotics in children because of cartilage and joint toxicity. So too was the off-label prescription of 1% xylometazoline nasal drops for babies: xylometazoline and oxymetazoline nasal drops are available in different drug concentrations specifically designed for different age groups, so the prescription to a baby of a concentration of drug designed for a child reflects poor care [157].

These and other histories teach numerous lessons individually. Considered as a group, however, they make it clear that safe and rational prescribing of drugs to children requires controlled studies of drugs in children. Concern regarding this issue was raised some years ago, as a result of which various changes were made to the regulations in the EU and in the United States to promote the generation of specific developments for the paediatric population. These measures and an assessment of how well they have worked are dealt with in further detail in Chapter 8.

Other patient populations

In the same way as there are specific risks to the child, there are also specific risks to other particular patient groups. We already know the risks to the pregnant woman, which is that of the risk to the unborn child. So when thalidomide was reintroduced for the treatment of leprosy and myeloma, a Risk Evaluation and Management Strategy, originally given the acronym STEPS, was introduced. Another example is letrozole (Femara™), approved as a breast cancer therapy for postmenopausal women but accompanied by a warning after it was found to be associated with a tripling of the normal rate of birth defects when used in pregnant women for fertility treatment. Despite this warning, some fertility doctors still prescribe the drug off-label to help women become pregnant.

There are other examples that involve secondary uses of drugs specifically in pregnant women but for which regulatory approval studies are slow in taking place. An example concerns misoprostol, used for the prevention of stomach ulcers. There is good evidence for the efficacy of misoprostol in obstetrics, to treat miscarriage, to induce labour and to induce abortion. However, the manufacturer, perhaps not wanting to be associated with an abortion drug, would not seek approval from the FDA for any reproductive health uses [158]. Publicly funded studies were eventually performed.

Nowadays all new drugs have been through teratogenicity testing to assess reproductive risk in animals. But that is not the same as testing in humans, and some older drugs have not been specifically assessed in this way. The lack of safety information in pregnancy makes off-label treatment much more likely and problematic in this group.

A similar issue applies for elderly patients, in some ways more so. Given the number of pills taken by this group of people, there is a more widespread danger than for any other group. Sometimes specific risks are associated with certain drug use in this population. The clotting (thrombogenic) risk with antipsychotics as used in dementia (see p. 10) is a case in point. Actually, the risks go wider than that. In elderly patients, the likely number of patients needed to be treated to induce the specific adverse event was calculated. This is the so-called number-needed-to-harm (NNH) value, and for the use of antipsychotics in elderly patients, the NNH values are as follows: increased risk of death (NNH = 87), stroke (NNH = 53 for risperidone), extrapyramidal symptoms (NNH = 10–20 for olanzapine, risperidone) and urinary tract symptoms (NNH range = 16–36) [159]. Again, these are remarkably hazardous drugs when administered in this off-label fashion.

Fatal ADRs

This brings us to the issue of fatal ADRs. How much does safety in drug use really matter, in terms of the ultimate arbiter of safety, or lack of it? We need to deal with another poorly appreciated matter; in other words, how often the drugs we rely on to save lives actually do the opposite and cause death. A study looking at ADRs in hospitalised patients in the United States from 1994 estimated that there were 106 000 deaths resulting from these ADRs [160]. This study, when published in 1998, attracted a great deal of attention; according to these results, ADRs are the fourth leading cause of death – ahead of pulmonary disease, diabetes, AIDS, pneumonia, accidents and automobile deaths. Fatal ADRs are therefore a surprisingly frequent occurrence. The obvious question is what is causing these fatalities, and could they be avoided? One immediate possibility is that of incorrect dosing being a cause; another is that of a prescription being written despite the drug being contraindicated. Both of these characteristics would qualify as off-label medicine. Preventable ADRs are a significant but heterogeneous component, probably composing 40–60% of the overall number [161]. Examples include too high a dose of an anti-hypertensive, giving rise to bradycardia or hypotension and gastric bleeding associated with the use of an NSAID; anti-infectives prescribed despite a history of allergy; overdose of an anticoagulant resulting in haemorrhage; and opiate overdose associated with respiratory depression.

Nobody has reported specifically on how many of the fatal ADRs are due to off-label medicine, but let's see if we can produce an estimate, using the information so far communicated. From page 71, adverse events run at around two- to threefold that of drugs prescribed on-label; we noted in that section that the proportion of serious events in off-label use is higher, but there are limited data for this factor, so (cautiously) let's exclude it. Doing so will probably give rise to a lower figure of off-label fatal ADRs rather than the opposite. Noting the previously quoted number of overall deaths associated with drug-related adverse events (106 000) and assuming a risk factor of between two- and threefold, we can calculate that between one in three and three in seven drug-related deaths occur as a result of this practice, namely, 35 000–45 000 per year in the United States. On a pro rata basis, using the figures quoted in the Introduction on p. xv, this equates to a figure of approximately 100 000 deaths annually across the OECD countries.

Can we approach a better understanding of this issue if we also look at it the other way around, in other words identifying off-label drug use and seeing how much of it is associated with fatal adverse events?

Let's take the example of quinine. A study from Eguale, looking at prescriptions written for off-label indications, reports that quinine is frequently prescribed for the treatment of night-time leg cramps. This is despite the assessed evidence that quinine derivatives are not recommended for routine use in this indication, being only modestly effective and hampered by their toxicity [162]. In 2010, the FDA placed a warning against the off-label use of quinine sulphate for nocturnal leg cramps, or restless leg syndrome, and a black box warning was added to the product label. Over a 3-year period to 2008, the FDA received 38 reports of serious side effects associated with the use of quinine including some cases of permanent kidney impairment and two deaths. Of these 38 ADRs, 37 were related to quinine being used for treatment outside malaria, which is the only FDA-approved use of the drug. In Canada, a similar situation applied, with 71 ADRs, but in only 4 cases where it was prescribed for malaria. Thus, the vast majority of fatal ADRs associated with quinine use are for off-label purposes.

Let's take another example, moxifloxacin. A 2003 study funded by the National Institutes of Health reviewed 100 emergency room prescriptions for fluoroquinolones and found that only 19 were written for appropriate conditions and only 1 was given in the correct dose and for the proper duration [163]. Most of the off-label prescriptions were judged inappropriate because there was another agent that was a preferred, first-line alternative. Of the 19 patients who received a fluoroquinolone for an appropriate indication, only 1 received both the correct dose and duration of therapy.

In addition to these two examples, there are various others which I have referred to elsewhere in this book. They include the excess thrombogenesis associated with the use of antipsychotics in elderly demented patients (p. 10); the use of erythropoietin for off-label indications, in which thrombogenesis is associated with higher doses (p. 58); the use of BMP-2 for off-label reasons (p. 15); or the use of factor VIIa for non-approved uses (p. 13).

I am also concerned that these figures underestimate the problem because of the way ADRs are reported. Despite the fact that off-label medicine is legal, there may be a reluctance to report, even in an anonymised way, deaths thought to be associated with off-label or unlicensed drug usage. A similar conclusion was reached by an analysis of fatal ADRs in paediatric medicine [164]. The reluctance may be related to the possible liability of the doctor in such situations, as we shall see in Chapter 6. But beyond the sheer number of excess deaths associated with off-label medicine that I estimated earlier, this is more than a safety issue, since we are dealing with a practice in medicine which is, by any acceptable standard, anything but evidence based.

Quality of evidence

The fallacy of the argument for use of a drug outside regulatory approval is widely appreciated when the evidence for its efficacy is poor. Off-label medicine can only really be appropriate in these circumstances if the patient is seriously ill and there is no alternative treatment. Where evidence is poor, the safety of the drug also needs to be concomitantly greater. This can be easily seen by the invalidity of a prescription of, for instance, an antibiotic for the common cold or in fact of any ineffective therapeutic treatment (like homeopathy).

'If there is not evidence presented to the FDA about a given indication, it certainly is a user-beware situation', said Jerry Avorn, professor of medicine at Harvard Medical School. A similar view comes from Adriane Fugh-Berman, MD, associate professor of Georgetown University Medical Center and director of PharmedOut. org, who says that 'Without clinical trials, physicians are essentially operating in an "evidence-free zone"'. PharmedOut is a project that educates doctors about the influence of pharmaceutical companies on prescribing patterns. Another quote comes from Diana Zuckerman, president of the National Research Center for Women and Families in Washington: 'So many drugs are being used by so many patients without objective research [or] evidence, based on their experience with relatively small numbers of patients, doctors may believe that the drugs are effective, but that is not conclusive evidence – and certainly not evidence that those medications are the best choices available' [165].

Strong evidence

I want to move on now to talk a little about situations where there is good evidence for off-label medicine. There are two reasons for this. The first is to balance the perception that we are confronted by quackery everywhere we look, since the landscape is actually quite heterogeneous, with small pockets of strongly evidenced practice. The second reason is to see how the good evidence was produced, outside a model of development dominated by industry support.

We have mentioned previously in Chapter 2 that tricyclic antidepressants are often used for the treatment of certain forms of neuropathic pain, yet they do not have FDA approval for this use. The evidence for these drugs, in particular amitriptyline, is substantial, and despite the lack of regulatory sanction, this class of drugs is considered a first-line therapeutic option.

The use of aspirin provides another interesting example where off-label use is a standard treatment. Aspirin was widely used before the establishment of the regulatory systems in all developed countries. In the United States, for instance, it was approved as an existing drug without the rigorous testing that modern medications undergo (in US parlance, 'grandfathered'). Currently, aspirin is approved for its analgesic effects (such as in patients with pain, fever, rheumatic diseases) as well as for its cardiovascular thrombolytic effects (e.g. in acute myocardial infarction, angina pectoris, coronary artery bypass grafting and previous cerebrovascular disease or myocardial infarction). There are some closely related indications where aspirin does not have an indication, yet guidelines recommend its use in these patients; an example is the use for prevention of coronary disease in diabetic patients. Therefore, aspirin prophylaxis for coronary disease in high-risk patients is an off-label use.

The journey that aspirin has taken to get to its current position is a long one, and still incomplete. Beyond the thrombolytic effect, its anti-cancer potential is still being delineated (see Chapter 2). β-Blockers are another interesting group of compounds whose role in therapy has evolved over a 50-year time span since their first discovery by Sir James Black in 1962. Their first approved use was in the treatment of tachyarrhythmia (an abnormally rapid and irregular heart rate). They subsequently became primarily known for their effects in hypertension and angina, after being administered to patients with arrhythmia and one or other of these conditions, but initially this was considered off-label. Once they became established anti-hypertensive and anti-anginal agents, they were then considered for other uses too. One of these areas where they are now known to have important benefits is in the treatment of congestive heart failure, where they are now the standard of care for nearly all patients. However, in early trials, these drugs actually made the condition worse in the short run, and it was only after prolonged treatment that patients improved. The historical basis for using β-blockers in heart failure dates back to 1973, but until 1985, there were only small short-term trials. Then during the subsequent clinical examination, it became clear that they held possibilities in this condition; by 1996, they were considered 'plausible' agents. It therefore took 23 years for this well-known group of compounds to change from being considered unacceptable to potentially useful. A further decade ensued before large-scale studies could be completed, leading ultimately to the establishment of regulatory approval for three compounds: carvedilol, nebivolol and metoprolol.

Another thing that happened in these 50 years was that the heterogeneity among these compounds became evident. Initially, they were known as β-blockers to differentiate them from compounds that blocked the alpha receptor for noradrenaline. Once the first agent was approved, pharmaceutical companies clamoured to reproduce its success, and a group of over 40 drugs became collectively known as 'β-blockers'. But over time, the β-adrenergic receptor itself became divided into β-1, β-2, and β-3, and the former homogeneity among the β-blocker class was established to be simplistic.

In hypertension, the differences are minor, but the pharmacological heterogeneity among the group can manifest itself in important ways in other conditions. Beyond heart failure, β-blockers have been shown to have beneficial effects in cachexia, the wasting disease most commonly associated with cancer. This research followed shortly after the long-term trials in heart failure, since cachexia also accompanies heart failure in about 10% of patients. By following the weight of these patients, it was found that certain β-blockers were able to prevent the loss of weight that is characteristic of cachexia [166]. But it became more interesting when the whole β-blocker class were investigated in comparison with one another. From an *in vivo* experimental analysis, it became clear that the combination of β-1 blockade, β-2 partial stimulation and a third component, 5-HT1A receptor antagonism, produced the most beneficial effect on cachexia. A clinical development employing the new drug MT-102 (based on the s-isomer of pindolol) has recently established a profound beneficial clinical effect in mid-stage trials in cancer-associated cachexia. I feel particularly privileged to have been associated with this discovery, along with the internationally renowned cardiologists Stefan Anker and Andrew Coats, and am the named inventor on the patent [167].

In another example, researchers discovered in the 1970s that the NSAID indomethacin was efficacious for closing a persistent, symptomatic patent ductus arteriosus (hole in the heart) in newborns. Thus, indomethacin became the treatment of choice for many affected newborns in an attempt to avoid curative surgery, though it never received formal regulatory approval for this indication [168]. As another example, SSRI antidepressants (e.g. paroxetine, sertraline and fluoxetine) and the tricyclic antidepressant clomipramine increase ejaculatory control and delay ejaculation in men with premature ejaculation, a use that has been proposed by the American Urological Association, based on significant clinical evidence, but not yet received regulatory clearance [169]. Definitive opinions in these diverse areas is difficult to obtain, but absent the incursion of the regulators, the UK National Institute for Health and Care Excellence has started to review individual examples in this area, as discussed further on p. 120.

In summary, there is a wide spectrum of different levels of evidence behind off-label uses of existing drugs for new indications, outside their approved use. Although the bulk of off-label prescriptions are written with little or no good scientific evidence, in some cases, the utility is well established and may even be standard of care. Outside industry-backed clinical development, there are relatively few routes by which this evidence in support of new indications can be gathered, so they are much slower to become mainstream, leaving patients waiting many more years without large-scale risk–benefit assessments being made. This is a big problem, for which I will attempt to offer solutions in Chapter 8.

Poor evidence

The history of medicine began with herbal remedies, which are notorious for their proposed utility in an implausibly long list of diseases. The same vagueness as to what it should be used for makes much alternative medicine suspect, like the ads on the front of the acupuncture shop in Cambridge, England (Figure 5.2). I particularly like the fact that at the end of a long list of unrelated diseases, the shop advertises its ability to treat a final medical condition, which it calls 'ETC'. Do you think, therefore, it also works as a treatment for curing a condition of 'excessive gullibility'?

Here's an example of something at this end of the evidential spectrum. Chelation therapy is a legitimate FDA-approved therapy for heavy metal poisoning. It uses either oral or intravenous drugs that bind to heavy metals and make them easier to

Figure 5.2 Really?

excrete. The treatment is regulatorily approved for lead poisoning, hypercalcaemia, and digitalis toxicity, where chelation literally binds the offending foreign chemical and assists in its removal from the body. However, chelation therapy has also been widely used off-label for decades to treat a long list of diseases and disorders. The evidence for many of these diseases (like autism and dystonia) is more or less non-existent [170], yet one of its proponents, Dr H. Ray Evers, argued successfully in 1978 for his right as a physician to prescribe off-label medication. Dr Evers was one of 100 physicians who have administered chelation therapy for inappropriate reasons. During his career, he claimed to have treated more than 20 000 patients and supervised more than 500 000 chelation treatments, resulting in at least 14 patient deaths. Fortunately, the forces of logic ultimately prevailed, and in 1986, Dr Evers had his licence to practice medicine revoked for gross malpractice.

A notable lack of evidence accompanies other more mainstream off-label medicine. A survey of 150 million off-label prescriptions in the United States by Radley, Finkelstein and Stafford found that 73% had little or no scientific support, even when sources other than the product information were searched. Thus, they found only a small proportion of off-label prescribing may be justified by scientific evidence [9]. A 2012 study from McGill University in Canada found a similar proportion from analysing the electronic health records of over a quarter of a million electronic prescriptions, reporting that 79% of off-label prescriptions lacked strong scientific evidence [43].

Healthcare watchdogs are therefore concerned that some drugs are prescribed off-label without conclusive testing for effectiveness and safety. In an effort to understand the situation more thoroughly, an organisation called the Developing Evidence to Inform Decisions about Effectiveness Network, which is part of the Agency for Healthcare Research and Quality (a division of the US government), commissioned a report from some experts in this area, including Randall Stafford, who co-authored the Radley et al. paper from 2006 [171]. They looked at what drugs were associated with these off-label prescriptions and what diseases or conditions were being treated, over the period from 2005 to 2007. Then, they looked to see what evidence was available for the indications, developing a traffic-light form of stratification, in which drug–indication pairs were labelled red, yellow or green, representing inadequate, uncertain and good levels of evidence. The products with inadequate evidence are listed in Table 5.2, ranked by frequency of prescription. The data from their study highlight a group of drugs as priorities for further study, where the extent of non-evidence-based prescription was particularly high. These were some of the most commonly prescribed drugs from the time of the study. Nowadays, the order may have changed slightly, but the central message is the same: many of the most widely used drugs are substantially prescribed with little evidence to back their use. This is a grave scandal at the heart of our healthcare system, an offence to the principle of evidence-based medicine.

How can this happen? Well, a partial answer to this question, and an issue that has recently been highlighted by the AllTrials group (http://www.alltrials.net), is that positive bias in scientific literature can convey a mistakenly positive opinion of a therapy which is not borne out by the facts. Medical professionals then become confused about what the real evidence is and where the risk–benefit balance properly lies. Lacking the compass of a regulatory view, they have to rely on what they hear from pharmaceutical company marketing campaigns without a balanced or thorough assessment.

There is a high proportion of antidepressants and antipsychotics in Table 5.2, like quetiapine, risperidone, escitalopram and so on. I have talked previously about the

Table 5.2 Drug use, by on- and off-label status and level of supporting evidence, ranked by decreasing order of off-label uses with inadequate evidence.

Rank	Drug	Therapeutic class	Prescriptions (%)		Good evidence (%)	Uncertain evidence (%)	Inadequate evidence (%)	Prescriptions with inadequate evidence (000s)
			On-label (%)	Off-label (%)				
1	Quetiapine	Antipsychotic	25	75	0	0	100	6507
2	Warfarin	Anticoagulant	71	29	0	0	100	5325
3	Clonazepam	Benzodiazepine	23	77	0	0	100	4235
4	Escitalopram	Antidepressant	86	14	0	0	100	3580
5	Gabapentin	Anti-epileptic	7	93	1	54	45	2827
6	Promethazine	Antihistamine	7	93	0	0	100	2732
7	Risperidone	Antipsychotic	37	63	0	36	64	2569
8	Digoxin	Heart failure	65	35	0	0	100	2561
9	Lorazepam	Benzodiazepine	61	39	0	0	100	2490
10	Lisinopril	Anti-hypertensive	93	7	9	0	91	2374

Reprinted with permission from Ref. [45]. © Nature Publishing Group/Macmillan.

use of antipsychotics for behavioural problems in dementia and PTSD. Some other uses promoted by companies include anxiety, obsessive–compulsive disorder, eating disorders, insomnia, personality disorders, depression and substance abuse. Despite some evidence for the benefit of individual drugs in some of these conditions, the evidence is too weak for their use in these conditions to be characterised as 'evidence based', and as we know, there are significant safety problems too [159]. This is really an extension of the point made in the last chapter, in which pharmaceutical companies market their products for off-label indications and bypass the regulatory system.

However, this is only part of the story. Some of the drugs included in Table 5.2 are old generic products, like warfarin, promethazine, lorazepam and so on. These drugs have not been actively promoted for years. I will talk about this more in the last chapter, where I propose some solutions to the problems we face, but I think we need a more holistic set of answers than just more regulations that the industry will find their way around. In the meantime, I want to address another issue that was identified in a German study on paediatric off-label medicine, where it was suggested that a significant component reason for the prevalence of the practice might be poor medical professional knowledge [7].

Doctors do not know evidence

Given the complexity and difficulty of knowing the evidence behind every possible therapeutic agent potentially utilisable for any particular patient, clinicians need comprehensive sources of reliable information in order to make prescription decisions. Fortunately, such compendia are available and considered standard reference works in the countries for which they are written. Thus, in the United Kingdom, we have the British National Formulary (BNF); in the United States, it is the Physician's Desk Reference; in Germany, the 'Gelbe Liste' and 'Rote Liste'; and in Australia, the Australian Medicines Handbook. Also in the United Kingdom, a BNF for Children has been specifically established, which presents essential practical information to help healthcare professionals prescribe, monitor, supply and administer medicines for childhood disorders. These books have been standards for decades: the Physician Desk Reference, for instance, has been the authoritative source on prescription drugs in the United States for 68 years. In addition to the availability of these reference works in hard copy, doctors also have access to information in electronic form, with sites like rxlist.com and drugs.com (as well as many others) providing free, up to date label information on all prescription drugs. Many practices also have bespoke software to guide their professionals' prescription choices, and manufacturers also provide smartphone apps so that doctors can have access to the most authoritative information on the move. This is both important and useful since the prescriber needs to know what is considered 'standard of care' not just for the ethical reasons in relation to his or her patients, but also for legal ones. As I shall outline in more detail later (see Chapter 6), there is a legal liability consequence of off-label prescription.

Despite the availability of all of these information sources and support tools, there is concern about the level of knowledge of some doctors in some areas. In a 2009 study of nearly 600 primary care physicians and 600 psychiatrists in the United States, the average respondent incorrectly identified nearly half of the drug–indication pairs (i.e. whether a particular drug should be prescribed for a particular condition) of the survey. This mistaken belief could encourage them to prescribe these drugs, despite the lack of scientific evidence supporting such use. Over two-fifths of physicians believed at least one drug–indication pair with uncertain or no supporting evidence (e.g. quetiapine [Seroquel™] for dementia with agitation) was FDA approved [34].

Furthermore, some physicians surveyed erroneously believed that drugs were FDA approved for specific conditions despite prior 'black box' warnings, indicating increased adverse events, including fatal ones, in the population being treated.

When the FDA warnings put new warnings in place, most physicians often do not follow them: a survey of health care professionals specializing in geriatrics found that although most were aware, less than half of prescribers reported that they changed their prescribing habits based on the notification [172]. Repeated examples solidify the case: as we saw in the use of quinine outside malaria (see p. 88) or letrozole in fertility treatment (see p. 86). Similarly, patients suffered when the macrolide antibiotic telithromycin was prescribed in various non-serious infectious conditions, after physicians ignored FDA warnings about severe liver injury and advice that it should be used only for pneumonia [173].

'Some physicians and healthcare experts maintain that physicians should know the evidence, not the FDA labelling. However, knowledge about FDA labelling can be important because FDA approval of a drug for a specific indication indicates a clear threshold of evidence supporting that use', said Donna Chen, MD, assistant professor of biomedical ethics, public health sciences, and psychiatry at the University of Virginia. Knowledge of the regulatory status is therefore a shortcut to knowing the evidence, which, with today's ever-expanding literature base, is impossible to know fully. Moreover, unlike other sources of unbiased information, a key aspect of a regulatory decision is that of a balance of risk and efficacy, not just efficacy alone.

There are two arguments medical professionals may use in their defence. The first is that the subtlety of real medicine often involves the need to prescribe to patients who are very near the definition on the label, but formally outside it. In the discussion around diagnosis, in Chapter 3, I touched on the complexity of diagnosing mental health disorders, with overlapping categorisations and subcategorisations. This is clearly an area of difficulty for the physician, but we allow for it, giving doctors latitude in the way they prescribe. Regulatory labels have been defined quite precisely to match up with the available evidence. The rest of the health system, from patient to manufacturer to reimbursement authority, is bound by this kind of precision (in the lawsuits mounted by the FDA against pharmaceutical companies for off-label promotion, there is a demand for an even more stringent requirement for abidance to the precise labelling). Medical professionals have the freedom to prescribe according to patient needs outside the framework imposed by the product label, but that is not the same as allowing the doctor to ignore what the framework says.

Second, it may be argued that it is sometimes difficult for prescribers to correctly identify what is real and what is promotional. Physicians at medical meetings may not realise that a fellow practitioner vouching for an off-label use has been hired by the pharmaceutical company to give the presentation. While the opportunity for off-label medicine is regarded as a necessary professional freedom, it also poses a difficult terrain that is almost impossible to know well, and pharmaceutical companies can colour this landscape with their own design. Even so, I am doubtful that this argument can be taken very far. Not only is the regulatory label printed and included in the pharmaceutical package, but it is available through reference books, electronic formats and internet searches, either in the office, at home or on the move. Medical professionals are a highly intelligent and somewhat sceptical clientele which represent a group not likely to be persuaded by the simple clarion call of an advertising jingle. They have their own ways of simplifying this jungle of information, whereby relationships between drugs and between indications guide the choices they make for their patients. It is incorrect to justify this behaviour as something practitioners inadvertently slip into, bemused by a complex array of intertwining facts that needs to be carried in their heads. Instead, off-label medication is actually something that many physicians actively seek out as part of their attitudes to therapy, regardless of whether the idea is suggested to them by pharmaceutical company promotion. Presented with a problem patient, they look for relationships to other patients, other conditions or other drugs.

Proximity of off-label to on-label

Based on indication

According to the relatedness principle, a doctor will expect that if a drug works in one indication, the same drug will work in a related indication; or alternatively, a related drug will work in the same indication.

Let's take the first case. Off-label use may result when drugs are used for clinically related conditions with overlapping features, such as the physiologically linked diseases of type II diabetes and polycystic ovarian syndrome (PCOS). Both of these conditions involve sub-sensitivity to insulin, and metformin is a well-accepted anti-diabetic treatment, so its use in PCOS is a valid medical hypothesis. One of the problems in PCOS is that women have lower than normal levels of fertility, so an indicative outcome from successful treatment is an increase in live births for such patients. However, results for the use of metformin in this condition are mixed, with some evidence for an increase in ovulations, but a failure for this to translate into motherhood. Metformin therefore remains an off-label experimental treatment in PCOS, and the regulatory label is restricted to diabetes [174]. Similarly, asthma treatments containing corticosteroids have been used for chronic obstructive pulmonary disease (COPD). But despite the theory and the fact that inhaled corticosteroids are 'gold standard' therapy for asthma, experts such as Peter Barnes are sceptical that they have much benefit in COPD, and in this condition, they are accompanied by increased risks of pneumonia and detrimental effects on bones [175].

Off-label use is particularly widespread in oncology, partly because of the large number of cancer subtypes, but also because of the difficulties involved in performing clinical trials in rare cancers, the rapid diffusion of preliminary results and delays in the approval of new drugs by regulatory agencies. All of this means that regulatory approval lags behind the ways that drugs are used in cancer, and they are not labelled for all of the indications for which they are thought to be effectively employed. Regulatory labelling is usually very precise in terms of type of tumour, association, line and schedule of treatment. As a result, the American Cancer Society has reported that up to half of the prescriptions in cancer are written off-label. Other studies report slightly different values, with the US General Accounting Office suggesting a figure of 33% in 1991 [176]; within this overall figure, there are hot spots – for instance, the single anti-cancer agent rituximab has been suggested to be used off-label 75% of the time [177]. Many of the main chemotherapeutic agents are used widely across all different types of cancer. But is there a presumption that anti-cancer agents can be used in situations where they really can't?

When bevacizumab (Avastin™), an inhibitor of blood vessel growth, was approved in 2004 for colorectal cancer, there was a suspicion it might be far more widely useful than that. Blood vessel growth (angiogenesis) is a general prerequisite for the growth of solid tumours. In fact, Avastin is now also approved for lung, kidney and ovarian cancers. Given this record, shouldn't it also be useful for breast cancer? Well, the FDA apparently thought so, allowing Avastin 'accelerated approval' in 2008, after preliminary studies showed that it helped slow tumour growth in this disease. But later experience negated this initial optimism, with follow-up studies finding that breast cancer patients who took it had no overall improvement in survival and suffered more serious side effects, such as blood clots and high blood pressure. In December 2010, the FDA decided to withdraw the approval of Avastin for breast cancer. In the United Kingdom, NICE followed suit, denying coverage of breast cancer by the NHS. (Avastin is also an example of a cancer drug that has proven to be of benefit in the metastatic setting but failed to improve outcomes in the adjuvant setting; adjuvant therapy for cancer is any treatment given after primary therapy to increase the chance of long-term survival. Despite this failure, some clinicians still use it for this reason [178].)

The Avastin in breast cancer case counters the view that drugs that are effective in indication A can always be successfully used in indication A-prime. However, there are as many exceptions to this as examples that follow the rule: policy is unpredictable and needs to be defined by evidence, an evidence that is gathered on a wide scale. Perversely, it is only because of the regulatory supervision that came with approval in 2008 that we are able to collect the post-approval statistics on the use of Avastin in breast cancer. Absent such approval, the off-label use of drugs in oncology, as well as in other areas, has been associated with a slow dissemination of information. We don't have the widespread independent evidence base upon which reliable decisions can be made, just individual doctors relying on their own patient experiences.

Clinicians' belief in the connected nature of different indications gives the basis for the phenomenon of 'therapeutic creep' (or 'indication creep'). This is where a drug licensed for one indication gradually becomes used for a closely related use, despite no clear evidence pointing out benefit in the latter use. Practitioners may extend the use of a treatment with real or suggestive therapeutic effects observed in a certain age group or in patients with a certain disease phenotype to other patients in whom the efficacy has never been demonstrated. We saw with the example of factor VIIa how this happened with an anticoagulant: initially approved for a rare form of haemophilia, it was extended into all kinds of conditions involving excess bleeding where later analysis showed that it didn't work and caused patient harm (see p. 13). We saw also how it was assumed, wrongly, that SSRIs would be useful in teenage depression because it was useful in adult depression.

Therapeutic creep affects the way in which testosterone therapy is used in situations not involving clear hypogonadism (its approved basis). Research shows an estimated 3–7% of middle-aged and older men have this testosterone deficiency, which they can be born with or develop after an injury or infection; testosterone deficiency causes poor libido, minimal muscles and scant body hair. However, men with any of these 'deficits' are not necessarily deficient in testosterone; it could be due to many other reasons. Testosterone therapy is nevertheless widely prescribed, for men without problems due to hypogonadism, and US sales of testosterone products grew enormously from about $18 million in 1988 to $1.6 billion in 2011. Patients without hypogonadism are medicated with the general aim of improving bone density, muscle mass, body composition and erectile dysfunction, though there is no good evidence of benefit in such patients. Some opinion leaders remark that testosterone level is the 'dipstick' of a man's health. Yet testosterone causes a significant side effect burden, with risks of the cardiovascular, respiratory and dermatological kind, as well as prostate hypertrophy. The growth of the 'low-T' or 'low-testosterone' movement has sparked a substantial controversy in terms of over-medication of normal men's health [179].

Another example involves the use of agents for gastro-oesophageal reflux disease (GERD) for patients with asthma. Children with asthma report symptoms of GERD more often than those without asthma, yet the association is poorly characterised. However, children with asthma but still without symptoms of GERD may sometimes be prescribed anti-GERD drugs such as the proton-pump inhibitors (e.g. omeprazole or lansoprazole). The use of anti-GERD drugs in child asthmatics is up to eightfold higher than comparator groups. Without proven benefit, the danger of this practice is the increased side effect burden consequent upon the proton-pump inhibitors like omeprazole, including increased risk of bone fracture [180].

'Therapeutic creep' is manna from heaven for a pharmaceutical marketing team, since clinicians can propel it and the company avoid regulatory censure by acting in pure response mode, driving sales by 'Continuing Medical Education' courses and cherry-picked marketing trials. But the difficulty of predicting the veracity of medical hypotheses beyond small-scale trials, as laid out in the earlier examples, makes it unwise to base medical practice on mere suppositions. In an effort to introduce more rigour into prescription decisions, recent contributors to this subject have proposed off-label medicine to be divided into 'supported', suppositional and investigational cases, on the basis that if prescriptions lack evidential basis, it is patient care that suffers [181].

Based on class of drug

Class effects can be observed among drugs with similar biological mechanisms of action. For example, the angiotensin-converting enzyme inhibitors are used for four potential cardiovascular indications, namely, hypertension, heart failure, post-myo-cardial infarction and diabetic kidney disease (nephropathy). Of the 10 angiotensin-converting enzyme inhibitors on the US market, only captopril has all of these listed as FDA approved in its label, although there is good evidence of a class-wide effect for angiotensin-converting enzyme inhibitors [45]. Similarly, rosuvastatin (Crestor™) is the only statin that is approved by the FDA for some patients with high C-reactive protein and normal low-density lipoprotein cholesterol levels, but others in the class may also work [49].

Despite these examples, the regulatory approval process treats each drug sepa-rately, consistent with the substantial within-class differences in safety and efficacy that are often seen among drugs with similar mechanistic profiles. A common biological mechanism does not always translate into efficacy of a single agent or group of agents in multiple indications. A good counter example to the 'class effect' is in the area of anti-inflammatory therapy, where the tumour necrosis factor (TNF) blockers such as etanercept (Enbrel™) and infliximab are currently approved for the treatment of the diseases shown in Table 5.3. Most of these are based around arthritis and gastrointestinal inflammation, in which there is good, though not complete, sim-ilarity between the indications for which etanercept (a TNF-receptor antagonist) and infliximab (a TNF antibody) are approved.

In addition to the diseases for which this class of agent has been approved, TNF has been found to mediate a range of other inflammatory diseases as shown in Table 5.4. In many of these conditions, either the blockade of TNF has been shown to

Table 5.3 Approved indications for TNF blockers.

Disease	Etanercept	Infliximab
Rheumatoid arthritis	Yes	Yes
Psoriatic arthritis	Yes	Yes
Ankylosing arthritis	Yes	Yes
Juvenile rheumatoid arthritis	Yes	Under investigation
Crohn's disease	No	Yes
Ulcerative colitis	No	Yes

Table 5.4 Indications in whose pathophysiology TNF has been implicated, but for which TNF blockers are not clinically effective.

Indication	Drug	References
Systemic inflammatory response syndrome	Infliximab/etanercept	187
Multiple sclerosis	Lenercept	188
Systemic vasculitis	Infliximab	189
Sjogren's syndrome	Infliximab	190
Wegener's granulomatosis	Etanercept	191
Autoimmune inner ear disease	Etanercept	192
COPD	Infliximab	186
Giant cell arteritis	Infliximab	193
Polymyalgia rheumatica	Infliximab	193
Chronic heart failure	Infliximab	194
Pulmonary sarcoidosis	Etanercept	195

be beneficial in preclinical models of the disease or TNF has been found to be over-produced in the human condition.

For instance, in COPD, the evidence base for a benefit of TNF blockade is founded on increased production of this biological protein in the sputum of COPD sufferers, both natively [182] and during exacerbations [183]; mice that have been genetically engineered to lack the receptor for TNF having substantially less cigarette smoke-induced emphysema [184]; and infliximab protecting against pulmonary emphysema in rats passively exposed to cigarette smoke [185]. But this evidence base from animal experiments does not translate into clinical efficacy: in a multicentre, randomised, double-blind, placebo-controlled, parallel-group, dose-finding clinical trial in 232 subjects with COPD, there was no treatment benefit of infliximab in the primary or secondary endpoints [186].

In fact, in all of the diseases in Table 5.4, a lack of effect has been shown clinically by one or other of the marketed TNF blockers. Thus, although the disease pathway is an important factor in de novo drug discovery and can sometimes be a good reason for successful off-label medicine, the hypothesis can also founder when put to the test.

Despite the lack of evidence, clinicians still prescribe these agents in areas without adequate scientific evidence or regulatory approval. Examples include the use of infliximab in asthma, COPD or Behcet's disease. They also look to increase the dose above that approved by the regulatory authorities, for treatment of inflammatory conditions which are approved, such as Crohn's disease, psoriasis or rheumatoid arthritis. In many of these cases, pharmaceutical companies have not been guilty of over-promoting into off-label areas, and it is the reimbursement authorities that act as gatekeepers for this behaviour (see Chapter 6) [196].

We have seen how the description of a 'class effect' often misrepresents differences among the class which could have important patient-level benefits (or lack thereof) for a certain select agent within the class. We have referred to the way in which the β-blockers are considered a homogeneous group in hypertension or angina, whereas they show very heterogeneous effects in cachexia. As another example, losartan is one of a group of angiotensin antagonists approved for the treatment of hypertension, yet it alone is effective in the control of gout [197].

Lessons about heterogeneity in a class, this time the NSAIDs, can also be learned from an appendix to the story of Vioxx™ (see p. 54). We saw previously how Merck compared their drug against naproxen in terms of gastrointestinal risk.

While showing a reduction in this primary measure, they also picked up an increase in cardiovascular risk which they conveniently, but incorrectly, ascribed to a benefit of naproxen, similar to the blood-thinning effects of another NSAID, aspirin, rather than a harm attributable to rofecoxib. Convenient or not, we now know that the correct conclusion involves a dual effect, namely, that in an absolute sense, Vioxx increases heart attacks and naproxen reduces them. The protective effect of naproxen is unusual among the NSAIDs, as shown by a later 2006 study [198].

Vioxx was an early exemplar of a group of more selective inhibitors of the COX-2 subtype of cyclo-oxygenase, the theory being that inhibition of the COX-1 enzyme enhanced gastrointestinal risk. Even though gastrointestinal side effects seem to accompany all non-selective COX inhibitors, a similar class effect is not found for heart attacks. This is not a straightforward story where cardiovascular risk of all COX-2 inhibitors is higher than non-selective COX inhibitors. The true situation, as we now know it, is shown in Figure 5.3, where gastrointestinal risk does indeed seem to be associated with COX-1 activity, but greater cardiovascular risk is present with both selective and non-selective anti-inflammatory COX inhibitors, so there are other well-known drugs with 'worst of both worlds' risks. Among the non-selective COX inhibitors are ibuprofen and diclofenac, which have also been found more recently to have a similar cardiovascular risk to Vioxx from long-term use [199]. It is not just me saying that Vioxx is not all bad: Catherine DeAngelis, the editor of the *Journal of the American Medical Association* (that published the Nissen study that originally showed the increased cardiovascular risk of Vioxx), has also said that she 'believe[s] Vioxx should not have been taken off the market'.

What has been revealed by the Vioxx debacle is more than just a hazard associated with a new branded drug, marketed by a large pharmaceutical company. Ibuprofen and diclofenac have been available for decades and are widely available as generic products, so the conventional wisdom that we 'know' about their risks because of the longevity of use is not true; moreover, ibuprofen is available over the counter, without prescription, a category generally reserved for the safest of pharmaceuticals. It defies logic to regard Vioxx as so dangerous that it is withdrawn from use,

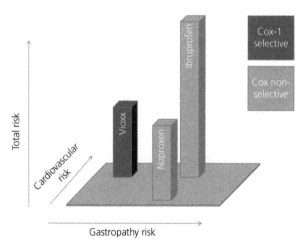

Figure 5.3 The COX-2 selective drug Vioxx reduced the gastropathy risk of the non-selective NSAIDs but substituted it with cardiovascular risk. Ibuprofen and naproxen, both non-selective NSAIDs, share similar gastropathy risks, but ibuprofen also carries a similar cardiovascular risk to Vioxx – the worst of both worlds.

even under a doctor's supervision, yet to continue to allow the sale of generic ibuprofen over the counter, when the latter commands the same cardiovascular risk and greater gastrointestinal risk.

While in no way wishing to diminish the cardiovascular part of this analysis, I would like to draw attention to the risks of gastropathy associated with the prescription of traditional NSAIDs, since these risks are huge and under-appreciated by the public (though not by many doctors) [200]. Clinical studies show that NSAID-related damage in the small intestine and colon accounts for between 13% and 40% of all serious gastrointestinal bleeding events. Major adverse gastrointestinal events attributed to NSAIDs are responsible for over 100 000 hospitalisations, US$2 billion in healthcare costs and 17 000 deaths in the United States each year, and this is just in regard to arthritis use (i.e. excluding other underlying reasons for needing a painkiller). These drugs have become so embedded in our pharmaceutical world that their dangers are assumed to be far less than they really are. In fact, NSAID gastropathy is the most common drug-related side effect of all, leading to more deaths annually than diseases like multiple myeloma, asthma, cervical cancer or Hodgkin's disease. It is rather perverse that the story of Vioxx, and its resultant withdrawal from the market for safety reasons, may indeed have led to patients switching back to cheap generic NSAIDs and their concomitantly greater dangers.

Debunking medical myths

In the eighteenth century, many sailors on long sea voyages succumbed to scurvy, a potentially fatal disease involving breakdown of connective tissue. At the time, there were various medical theories as to its cause. It was a serious problem that affected many of the seagoing population, and there were many hypotheses as to how to treat it. Diet was thought to be at its core, and possible curative dietary supplements included cider, sulphuric acid, vinegar, seawater and spicy paste with barley water. The Scottish physician James Lind believed that scurvy was caused by putrefaction of the body that could be cured with acids. He conducted the first ever clinical trial in 12 sailors suffering with the disease, to investigate all of these five medical treatments, plus one other, a daily ration of two oranges and a lemon. Alone of the treatments, the vitamin C-containing dietary supplement provided significant benefit, with one of the two treated sailors becoming fit for duty within 5 days and the other almost recovering by this time. Interestingly, Lind did not identify the central importance of vitamin C even after this experiment; he continued to believe similarly to most of the medical profession at the time, who contended that scurvy was essentially a result of ill-digested and putrefying food within the body, bad water, excessive work and a damp living environment which prevented healthful perspiration. Medical myths of that time often had data-defying lives of their own, but surely today, we could count on a more incisive scientific analysis.

Off-label medicine has sometimes been characterised as a long-standing medical practice that has, for commercial reasons, not been subjected to high-quality clinical trials and consequent regulatory scrutiny; however, according to this characterisation, that deficit should not detract from its rightful place in medical practice. Yet we have also seen how some long-established off-label uses have been shown to either be ineffective or harmful when subjected to proper evaluation (such as the deaths associated with chloramphenicol when used in paediatric care). Examples such as these are called 'medical reversals'; how common are they?

A recent study looked at 363 articles in *New England Journal of Medicine*, in particular analysing situations in which new studies contradict current practice [201]. The study found that only a minority (38%) of current practices were affirmed by this analysis, compared with 40% [148] that were shown to be ineffective and a further 22% where the evidence was inconclusive. This tells us the danger of accepting arguments about 'standard of care' without formal proof and of course raises some concern for the majority of off-label prescriptions, for which there is little or no scientific support.

A good example involves the use of bone marrow transplants in breast cancer. In 1986, doctors from the Harvard Medical School and other hallowed Bostonian medical institutions reported using very high doses of chemotherapy to eliminate residual cancer in patients with advanced disease and then resuscitating them with using their own bone marrow that had been previously harvested [202]. The results from a small, uncontrolled 'proof of principle' trial were positive, thus providing evidence of both feasibility and potential efficacy. The quality of the results (perhaps also combined with the eminence of the research institution from which they originated) persuaded others to adopt the approach, and it quickly became the 'standard of care'. Pressure built from patient advocacy groups for hospitals to provide this therapy to women and for insurance companies pay for it; in fact, some patients successfully sued their insurance companies when they were denied coverage for this form of treatment. Unfortunately, when randomised clinical trials were eventually performed some years later, the results showed that the procedure was no more effective and possibly more harmful than conventional therapy [203]. As a result, many thousands of women who were treated according to the so-called standard of care, between the clamour at the time of the first small-scale trials and the dull reality arising from the definitive set of studies, were in fact following a poorly evidenced medical practice that in fact did them little good.

Another example, which resulted in a major political scandal, occurred in France. Here, the drug Mediator™ (benfluorex) was licensed as an add-on for hyperlipidaemia and diabetes but was, until it was finally banned, routinely prescribed for general weight loss in non-diabetic patients. The French authorities estimate there were between 500 and 2000 resultant deaths during its 33 years on the market [204]. One should remember that the French public's opinion of doctors has not historically been a favourable one. Voltaire, the eighteenth-century French philosopher, is quoted as saying:

> Doctors are men who prescribe medicines of which they know little, to cure diseases of which they know less, in human beings of whom they know nothing.

That of course has since changed, since nowadays many doctors are women. However, even against this backdrop, the scandal shocked the French establishment and resulted in the resignation of Jean Marimbert, who was head of the French regulatory agency AFSSAPS. In January 2011, in an unusually strong form of disciplinary sanction, the French association of drug makers (LEEM) went as far as to suspend the manufacturer Servier, and the French health minister promised reform of the regulatory landscape.

This story began in the 1960s, when Servier was experimenting with derivatives of amphetamine as appetite suppressants. Early on, they identified compounds such as fenfluramine, dexfenfluramine and benfluorex (later marketed as Mediator). The first two of these compounds were differentiated from benfluorex, according to Servier, by their pharmacological effects and consequently by their medical use. Thus, the fenfluramines became marketed for their anorexigenic (appetite suppressant) effects for obesity, whereas benfluorex was approved by the French regulators in 1976 as an adjunct to diet for hyperlipidaemia and for diabetes plus obesity. Servier argued that benfluorex might act on lipid and cholesterol metabolism. However, in fact, all three compounds had similar effects on serotonin, causing increased serotonergic effects through a combination of increased release from nerve endings and reduced reuptake.

About 20 years later, as a result of the Fen–Phen scandal (see p. 79), fenfluramine and dexfenfluramine were found to be associated with cardiac valve defects (valvulopathy) and increased the risk of pulmonary hypertension. The toxicity was associated with increased serotonergic activity at the 5-HT2B receptor. The French regulators restricted the use of the fenfluramines but failed to spot the connection to benfluorex, because it was licensed for a different indication and predominantly prescribed outside regulatory sanction. However, the fact that benfluorex was widely prescribed off-label for its anorexigenic effects, rather than for its anti-diabetic and anti-hyperlipidaemic use, could not disguise its essential mechanistic similarity to the fenfluramine class of compound.

So this is a classic example of an accepted and widespread off-label use, which had not been subjected to a risk–benefit analysis in the treatment of obesity and continued to be used without this analysis for over 30 years. Early warnings of toxicity that appeared not just in France, but also in Italy and Spain, were ignored. Finally, research by Irène Frachon, a chest physician at the University of Brest, uncovered a number of people with valvular heart disease and showed their condition to be 17 times more likely if they had been treated with benfluorex [205]. This is a remarkably high ratio.

In summary, this chapter has revealed firstly that our expectations of safety for a product that has been approved in the context of a particular therapeutic use, patient group and dosing regimen, are not necessarily permissive for use outside those parameters. We have seen how, if approved in a serious condition, the safety profile is often not appropriate for a less serious condition; how, if approved for adults, it is not necessarily safe for children, expectant mothers or old people; and how a drug that's good for short-term use is not necessarily safe for longer-term use.

Then, in the consideration of efficacy, the quality of evidence that accompanies off-label medicine is mostly inadequate, because there are no defined standards. Doctors sometimes need the freedom to prescribe when patients present to them with serious, perhaps life-threatening conditions for which there are no approved agents. One can understand the difficulties imposed by those rare circumstances, when physicians must prioritise patient care above pharmaceutical regulations. However, I think most readers will be perplexed as to why this freedom to prescribe has been overused to the tune of one prescription in five. Regulations to require proven efficacy for approved treatments were put in place over a half century ago, precisely to protect patients. Unfortunately, not only are these regulations laid to one side far too often, but the outcomes from off-label medicine are not properly assessed and recorded. Myths develop about the utility of certain treatments which can become ingrained in medical practice without anyone ever undertaking formal studies to prove their validity. Far too often, when challenged and the facts are laid bare, do we then see the emperor has no clothes.

CHAPTER 6

Liability, injustice and reimbursement: who should pay?

A prescriber's ethical and professional duties

Before dealing with the legal issues around off-label prescribing, we need to put the ethical context right. The point has been made that off-label prescriptions should not be viewed as scientifically or ethically unsound when there are good clinical data to support a particular therapeutic indication [206]. The trouble is that there are only good data in about 25% of off-label use, suggesting that the other 75% *should* be regarded as ethically unsound. This difficulty has found its way into various professional medical guidelines as we saw earlier. The underlying scientific basis for off-label prescribing also becomes a significant legal issue in determining liability suits. The ethical context frames the legal situation, and given that the country-by-country situation differs with respect to professional guidelines, it is not surprising that there are differences between countries in terms of the liability issues.

In addition to the importance of the evidential basis, two other ethical factors come into play. First, the situation of the patient, and second that of alternative medical treatments. Clearly, the more serious the disease and the more bereft of licensed therapeutic alternatives, the more justifiable is an off-label prescription. The treatment of cancer in a child is perhaps one area highly likely to be seen as ethically justified for such a treatment option.

But there is one area where healthcare professionals are unlikely to get much sympathy, and that is when accepting the industry shilling for promulgating off-label pseudomedicine.

Medical professional participation in off-label promotion

We saw in Chapter 4 various examples of how the industry circumvented the regulatory hurdles in pursuit of an enlarged commercial opportunity. We shall see later in Chapter 7 the massive financial penalties that followed this activity, which covered various forms of off-label promotion, but also financial inducements for physicians to prescribe in this way. 'Financial inducements' are a polite term for bribery, and as anyone knows, there are two parties to this crime. I have more to say on the role of industry later, but we should not forget the role of the physician in this.

Despite the culpability of the large organisations, much of the wrongful behaviour centres on individuals, yet hardly any individuals from either the industry side or the physician side have been punished. For instance, in prosecuting GSK for off-label promotion of its antidepressant Paxil™ (paroxetine), it was revealed that the company had tried to win over doctors to prescribe off-market by paying for trips to Jamaica and Bermuda, as well as spa treatments, concerts and hunting excursions [207]. Responding to the concern, in 2013, the US government passed the Physician Payments Sunshine Act; 'the best disinfectant is sunshine' is a well-known quote

Off-label Prescribing – Justifying Unapproved Medicine, First Edition. David Cavalla.
© 2015 John Wiley & Sons, Ltd. Published 2015 by John Wiley & Sons, Ltd.

from US Supreme Court Justice Louis Brandeis, referring to the benefits of openness and transparency. Under the act, all payments above a minimal level must be reported as 'Open Payments'. The website ProPublica.org publishes the information on payments between industry and medical professionals, and this has had a significant effect in reducing direct transactions to doctors [208]. Also in 2013, GSK announced it would stop payments to physicians for promotional purposes. It is unclear at present whether GSK's lead will be followed by the rest of the industry, but nevertheless, most of the US pharmaceutical companies are now reporting reduced spending on physician speakers [209]. These moves are welcome in combatting excessive pharmaceutical marketing activities, even if not focussed solely on off-label promotion.

In addition to receiving payments for modifying their prescribing behaviour, the second way in which medical professionals may participate in promotional activity is as unwitting facilitators. Even though physicians are an intelligent group, and not easily persuaded just by pharmaceutical reps, a major problem with off-label prescriptions is that the evidence behind the new uses often originates from only one source. This source, being the manufacturer of the drug, is hardly likely to be independent. Other independent reviews of the information can be drowned by the corporate message, unlike the large amount of information and independent regulatory review to support the labelled, approved use. There are a whole host of methods by which the integrity of the medical profession is subjugated to the corporate intent. We are faced with issues like publication bias, ghost-written articles, small-scale studies and endpoints of dubious clinical merit to contend with.

The history of Neurontin™ is a case in point, where an in-depth analysis in 2013 countered the swathe of literature used to promote the product a decade earlier; the company-sponsored publications differed from internal company research reports which were revealed during the litigative process, suggesting the scientific publications were not unbiased and accurate descriptions of the actual trials that had been conducted [210]. In the absence of an authoritative counterview, resistance against this one-sided view of the evidentiary base requires much more time and in-depth analysis for which few (no?) medical professionals have the time [211]. If this practice is not an isolated event, we have a serious problem – biased publications leading to drugs becoming cemented in medical practice and used off-label without regulatory oversight – it is exactly to prevent this occurring that we have a regulatory system in the first place.

A prescriber's legal position

Consent

At the beginning of this book, I said that off-label medicine, 'for the most part, is all perfectly legal'. It is now time to unpick that statement more fully, since in some countries, there are significant restrictions on what prescriptions can be written for unapproved medicine. The first restriction concerns the disclosure to a patient that a treatment is off-label and the informed consent resulting therefrom.

For instance, in Italy, legislation explicitly highlights the importance of obtaining the patient's informed consent. Under the so-called 'La Bella' law from 1998 ('Legge di Bella'),

> …in single cases, the doctor may, under his own direct responsibility, and after informing the patient and obtaining his consent, use an industrial medicinal product for an indication or means of administration or form of administration or of use different to that authorized… if the doctor considers, on the basis of documented data, that the patient

cannot be treated with medicinal products for which there is an approved therapeutic indication or means or form of administration and provided that such use is known and in accordance with research published in internationally reputable scientific publications.

The story behind the law is interesting, since it originated from a proposal from a physiologist called Luigi di Bella in the late 1980s for a new anti-cancer treatment based on a combination of melatonin; bromocriptine; cabergoline; somatostatin; a mixture of retinoids; vitamins E, D and C; and cyclophosphamide. These ingredients were proposed to work synergistically, resulting in immunostimulant, anti-proliferative, antioxidant and anti-angiogenic effects. Nevertheless, the long list of ingredients makes it seem more like a herbal potion rather than a deterministically and scientifically derived therapeutic. This view is further solidified once we hear that it was also proposed to be non-toxic. A more remarkable advance in the treatment of cancer could hardly be imagined! But of course, the proposed benefits were ultimately shown to be entirely imaginary. A series of uncontrolled Phase II trials were conducted at public expense, but their results did not demonstrate any effect on tumour size reduction, and the Italian medical authorities declared the treatment to be without merit. Nevertheless, public pressure continued for the provision of somatostatin at public expense, as a result of which the government succumbed, but remained concerned about the detrimental effect on its finances. It wanted to avoid expending limited public funds on treatments that performed badly with respect to more efficacious comparators.

As a result, the 'La Bella' law was drafted. But after it was passed, a subsequent Italian Supreme Court decision clarified the situation in a very sensible fashion, requiring three conditions for off-label use to be valid: the informed consent of the patient had to be obtained; there should no on-label alternatives; and the off-label use must be known and in accordance with research published in internationally reputable scientific publications (to exclude a conference presentation or brief abstract, which might represent poorly substantiated results). Law 244 of 2007 (Legge Finanziaria 2008) goes even further by adding that the doctor's decision must be based on data resulting from Phase II clinical trials, which are the mid-stage of a compound's development where efficacy benefits have been demonstrated in small-scale trials in patients.

Under Italian law, a doctor may be held responsible if the patient both received insufficient information about the off-label treatment and is harmed as a result of that treatment. The Italian Supreme Court has addressed the criminal liability of doctors when they engage in off-label use without the informed consent of the patient. In a case involving the off-label paediatric prescription of Topamax™ (topiramate), for the treatment of obesity (whereas it was only approved for epilepsy), the doctor was found to have erred in not obtaining the informed consent of the parents.

In Spain, there is a body called the GENESIS group (Group for Innovation, Assessment, Standardization and Research in the selection of drugs of the Spanish Society of hospital pharmacy) which conducts evaluation studies on new indications for existing drugs. Recommendations are given by the group, and further approvals given by the hospital Pharmacy and Therapeutics Committees when strong evidence is available. A Royal Decree from 2009 authorises doctors to prescribe off-label in exceptional circumstances if there are no alternative authorised therapies for the particular patient, but the doctor is obliged to obtain prior informed consent from the patient.

In the United Kingdom, on the other hand, there is no legal requirement for a doctor to disclose to the patient when a drug prescription is unlicensed, but there are professional guidelines for obtaining consent (see p. 38), and off-label prescribing increases professional liability. There is also, under the Mental Care Act 2005, a specific requirement for a health professional to warn of potential dangers of a patient's treatment, which may apply in psychiatric medicine. Furthermore, despite the lack of general litigative power, the increasing drive for transparency in the United Kingdom has also resulted in the NHS guidelines that 'the person must be given full information about what the treatment involves, including the benefits and risks, whether there are reasonable alternative treatments, and what will happen if treatment does not go ahead. Healthcare professionals should not withhold information just because it may upset or unnerve the person'.

National requirements for informed consent vary across the EU, but it is generally agreed that physicians should inform their patients of the unlicensed nature of the proposed treatment, the reasons for proposing the treatment, any potential side effects, the risks and benefits and any available alternatives. The legal consequences of failure to obtain informed consent for an off-label treatment found their way into the French courts. In 2008, the Cour de Cassation (equivalent to the Supreme Court) assessed the liability of a doctor who had prescribed the off-label use of the vasodilator papaverine to treat the erectile dysfunction of a patient, resulting in total and irreversible erectile dysfunction. At that time, of course, Viagra™ would have been an approved alternative. The court held that the doctor was liable for harm suffered by the patient because he had failed to properly inform the patient about alternative treatments. The French health code says that doctors must seek the consent of their patients 'in all cases', and the judicial system has also held that compensation must be provided when informed consent is not obtained [212].

At a supranational level, informed consent is now a European right, and all EU member states are bound to protect it. The Council of Europe's Convention on the Protection of Human Rights and Dignity of the Human Being with regard to the Application of Biology and Medicine (Oviedo Convention) requires the informed consent of the patient for any medical treatment. The European Court of Human Rights has held that failure to protect the right to a 'free, express and informed' consent, as set out in Article 5 of the convention, constitutes a violation of the right to 'private life' of the European Convention on Human Rights, to which all EU member states are bound. Similarly, the Charter of Fundamental Rights of the European Union (European Charter) provides that in the field of medicine, 'the free and informed consent of the person concerned' must be respected, according to the procedures laid down by law. The requirement of informed consent under the Oviedo Convention and the European Charter applies to any medical treatment. For off-label use, informed consent means that the patient has adequate information about risks and benefit on both the appropriate approved and off-label treatments. For instance, in regard to off-label prescription of antipsychotics, the risks of stroke or even of cognitive impairment would presumably count as components of a discussion towards consent. EU regulations envisage off-label or unlicensed drug use only under limited circumstances involving 'special need'. These include authorised clinical trials, compassionate patient use when no other treatment is available, emergency scenarios (e.g. pandemics) or the discretion of a treating physician. Though this last clause may seem to offer a 'catch-all' exemption allowing the prescribing physician to use off-label medicine whenever he or she sees fit, there has actually been a legal decision to clarify what the EU regulations mean by 'special need', such that it should not apply when licensed alternatives are available (see p. 122).

The issue about the off-label nature of a prescription having legal implications for a prescriber is embodied in suggestions by the Canadian Medical Practitioners

Association (CMPA) that their members take the following series of actions to minimise the risk of liability when making an off-label prescription [213]:

- Consider if there is sufficient support from the medical literature (e.g. guidelines from medical specialty organisations) for the off-label use of the medication or product. Is the use in keeping with the present standards of practice?
- Document the rationale for using the medication or device off-label.
- Obtain a detailed history from patients and examine them to determine if they have a condition that would place them at increased risk of potential side effects from the off-label use of the drug or device.
- Obtain and document patients' consent after an appropriate discussion of the potential risks and benefits, and after a discussion about the medication or device being used in an off-label fashion.
- Document any questions asked by patients and the answers provided.
- Carefully monitor patients for side effects during or following on an off-label treatment.

The Canadian procedures are similar to (but more detailed than) those mandated in the United Kingdom by the GMC (see p. 38). However, there are no requirements in Canada to use a licensed alternative if that exists. The CMPA position is informed by their Supreme Court, which held that a physician should tell the patient of *any material risks* in relation to the treatment, such materiality being defined by the risks a reasonable patient would regard as such.

Relative to all these jurisdictions, the situation in the United States is far more lax. The courts there have not widely held that the doctrine of informed consent requires physicians to tell their patients when a drug is being prescribed for an off-label use. US law nevertheless enshrines the principle of informed consent; it's just that the disclosure doesn't have to include the regulatory status of the treatment. The law on informed consent goes back to the legal considerations around battery, and consists of unpermitted, unprivileged, intentional contact with another person. Even if there is no bodily harm, the intent of such harm can be enough. The patient's consent distinguishes permitted from unpermitted treatment. A claim of lack of informed consent usually accompanies an allegation of medical malpractice for wrongful diagnosis or treatment. The elements of the claim are generally that the physician did not present the risks and benefits of the proposed treatment and of alternative treatments; that with full information, the patient would have declined the treatment; and that the treatment, even though appropriate and carried out skilfully, was a substantial factor in causing the patient's injuries.

Lack of informed consent is an issue of self-determination. If a lack of informed consent can be demonstrated, it can reinforce a claim of medical malpractice or serve as an alternative point of attack when the case is otherwise weak. Informed consent requires a certain level of disclosure by the physician, which is based either on what a reasonable patient would want to hear or what a reasonable practitioner would provide. In a slight majority of US states, the 'reasonable practitioner' standard has been adopted, although a substantial minority have opted to adopt the 'reasonable patient' standard. Under the 'reasonable practitioner' standard, the tendency is to regard the relationship with a patient as 'doctor knows best' and to ignore the concept of patient autonomy to decide what is acceptable regardless of the type and likelihood of risks involved [214].

There have been proposals to extend the principle of informed consent in the United States to include the disclosure of the FDA regulatory position. Support for this position comes from an understanding of the attitudes of a 'reasonable patient' as evidenced by public polls on the issue of off-label drug use. In a 2006 poll, most of the public incorrectly believed that a drug could be prescribed only for its primary FDA-approved use, and a similar percentage felt that physicians should be prohibited from prescribing drugs for off-label use [68]. Although many courts do not require physicians to disclose the regulatory status, patients mostly have a different belief and concern regarding the use of unapproved medicine. One of the reasons why moves to tie regulatory status disclosure to informed consent have been stymied is because the US Congress has expressly forbidden the FDA from regulating the practice of medicine, and off-label use is considered the prerogative of the physician [84]. Counterviews to the public perception have also been expressed by industry lawyers, who oppose any attempt to require off-label disclosure by physicians to patients [91].

The US requirement for a physician to inform a patient about the FDA regulatory status of a treatment was tested in a case involving a bone screw device. Originally approved for use in long bones, the screws were also occasionally inserted by orthopaedic surgeons into patients' spines. In one of the lawsuits, *Klein v. Biscup* (Ohio, 1996), a lack of informed consent was alleged because the surgeon who had inserted the device had failed to tell the patient that it was not approved for use in her vertebra. After the surgery, the patient experienced pain in her legs as well as bladder and bowel control problems, and later she was diagnosed as having a fracture of her fifth lumbar vertebra. Her complaint regarding informed consent was unsuccessful, and in making its decision, the court noted that FDA regulatory labels were not intended to interfere with the practice of medicine; consequently, this fact was not 'medical information concerning a material risk'. The court found in favour of the defendant, since the surgeon had no legal duty to advise the patient of the FDA regulatory status of the device. Though not directly involving pharmaceuticals, this case is just one of several with similar findings. Despite there being no general requirement to disclose the off-label nature of a medicine, it is mandatory for a physician to disclose this if the drug or device to be used in an off-label manner is part of a clinical investigation or other experimental study.

With reference to the previous chapter on safety and efficacy, it might seem aberrant for the US courts to regard the FDA regulatory status as non-material with respect to patient risk. As I said, the basis for this comes from a stipulation placed by Congress on the powers of the FDA, such that it was expressly forbidden from regulating medicine. This is not the same, however, as saying that FDA information is irrelevant for medical practice. On the contrary, the FDA regulatory function is specifically mandated to imprint an acceptable standard of risk and benefit on a product as used for a particular condition and a particular set of patients. Physicians are not obliged to slavishly adopt the FDA commands; but the FDA approval status is used in the compendia, like the Physicians' Desk Reference, that physicians do adopt in their everyday prescription decisions. It is also included on the package insert that accompanies pharmaceutical products.

I see the relationship as similar to Government and the Legal Profession. Government does not regulate the practice of law, but it does codify the laws upon which legal decisions are based. Those decisions are up to legal professionals to make. Some commentators have suggested that FDA regulatory status is 'irrelevant' to the practice of medicine [215]. If that were so, how would physicians make their decisions in the best interests of patients? Are they able to somehow call upon a well of professional expertise that is devoid of regulatory input yet is still in the best interests of patients? No, of course not. Physicians are informed every day by FDA regulatory status but allowed to interpret that information in the best interests of their patients and to choose an off-label option if appropriate; in practice, that happens in around one in five instances. To say that FDA regulatory status is 'non-material' for patient risk is pure legal sophistry.

There is a better approach in such cases around the issue of informed consent. This would be to ask the medical professional defendant to show why the safety and efficacy of the off-label treatment is better than or the same as that of the approved treatment. When such a comparative assessment has been made, taking into account the much larger set of safety and efficacy information in an approved treatment, any difference in risk should be communicated to the patient. It would seem perfectly possible to adopt this approach without changing the essential function of the regulatory agencies. From the reasonable patient's point of view, any safety or efficacy assessment that differs from the regulatory standard would be material.

Liability

The second question related to the prescriber's legal position pertains to liability. Here, in most jurisdictions, the matter is more adverse to the physician. Thus, even though the off-label nature of a treatment may be merely a 'matter of medical judgement' and physicians may not, for example in the United States, be held liable for non-disclosure to patients, this may still be a contributory factor in assessing professional liability for medical negligence involving off-label treatments.

By prescribing a drug off-label, the physician is therefore taking a special responsibility. Formally, he is prescribing something which the regulatory body has not stated is safe and effective. The prescriber may therefore be asked to respond for any problem arising from the use of the drug as if it is outside the state of the art. It may be that this is not the case, but the burden of proof rests on the physician. Whether a given off-label prescription meets an acceptable standard of care will depend on the level of evidence available to support the use and how the prescriber used the available evidence.

Physicians have been frequently involved in legal claims due to an adverse reaction related to a medication prescribed for an off-label use. When a patient claims an injury from a prescription drug or medical device, he or she usually sues both the physician and the drug or device manufacturer. In such product liability lawsuits, plaintiffs' claims include those for negligence and/or strict liability against the product manufacturers for failure to warn of the potential risks of a particular drug or device.

In the United States, the legal basis for this involves an understanding of the principles of the 'learned intermediary'. Under this principle, the manufacturer only has to warn the physician about the risks involved – not the patient – and it is the physician's responsibility to pass these warnings on to the patient. The printed label which accompanies the product effectively does that. However, when the physician prescribes off-label, *the warnings cannot come from the manufacturer* since the product is not being sold for the unapproved, off-label purpose. This places a greater burden of liability on the physician's shoulders, up to and including some personal responsibility in such cases.

In a case in Maryland, *Robak v. Abbott Laboratories*, the liability of the pharmaceutical company was assessed with respect to the use of temafloxacin, a quinolone antibiotic (risks of which class were discussed on p. 75) for sinusitis, an off-label use. The court decided that the responsibility lay with the physician since the company could not have assessed the likely safety risks outside the labelled indication.

However, the outcome from product liability cases also depends on the extent to which the manufacturer knows about the off-label use. In a case in New Jersey, involving GSK's paroxetine (*Knipe v. SmithKline Beecham*), the fact that the company knew of the off-label use of the product in young patients made it responsible for warning physicians about the risk of suicidality associated with this use, and the court rebutted the attempt by the company to pass the responsibility for the damage on to the physician (as the 'learned intermediary'). In a case in Texas (*McNeil v. Wyeth*) involving the product Reglan™ (metoclopramide), prescribed for a period longer than 12 weeks, the patient developed tardive dyskinesia, a risk discussed earlier on p. 81. The court found that the company had a duty to warn physicians of the risks associated with this prolonged use because a majority of its sales came from these extended prescriptions.

Finally, in a case involving the off-label use of Avastin™ for eye conditions (*Weiser v. Genentech*), the plaintiff was blinded by a non-sterile formulation of the drug (which is approved for cancer) after being injected into the patient's eye. In this case, the product was reformulated by a compounding pharmacy but without the stringent sterility procedures involved in the manufacture of the approved product, Lucentis™. It is interesting that the plaintiff chose to pursue the manufacturer as well as the compounding pharmacy. The off-label use of Avastin in age-related macular degeneration (AMD) is described in further detail on p. 117. No outcome from this case, which started in 2013, has yet been reported.

Product liability also tends to be assessed as the responsibility of the off-label prescriber in the United Kingdom. Normally, if a pharmaceutical product is defective,[1] the patient can bring a liability suit against the licence holder or manufacturer without needing to prove negligence. This is called 'strict liability'. But in cases where the product is being used outside the label, the purchaser (i.e. healthcare provider) retains liability for the quality of the product and ensuing damages, with the manufacturer being considered a subcontractor. In combination with the GMC guidelines, UK doctors face much uncertainty about either disciplinary or litigative action against them for off-label prescriptions, and in situations like paediatric medicine, this poses a particularly difficult situation because so many drugs have not been tested in children [216].

In Germany too, pharmaceutical product liability in these situations generally lies with the physician. Off-label use is permitted for life-threatening conditions where there is no alternative therapeutic option with evidence of likely success. If there is no alternative therapeutic option, it is also permitted for serious disease provided the evidence is good, but if there is an alternative, off-label prescriptions are not permitted for minor ailments. German doctors therefore need to know both the regulatory status and the evidence behind a prospective off-label instance; doctors also have an increased duty to inform patients about the therapy and to monitor it. If a patient suffers from an off-label treatment, the physician is liable, potentially under both civil and criminal law; however, in most civil cases, this is covered by personal liability insurance [90,217].

Finally, as well as product liability suits, patients may also bring medical malpractice suits for negligence. In the United States, there have been some suits against the manufacturer for allowing off-label use to take place; these have not been generally successful. On the other hand, plaintiffs have been more successful in claims against physicians, so long as they can show the off-label use deviated from an acceptable and prevailing standard of practice. If physicians use a product for an indication not in the approved labelling, they have the responsibility to be well informed about the product, to base its use on a firm scientific rationale and on sound medical evidence and to maintain records of the product's use and effects. The principle is that off-label medicine is a matter for medical judgement, so a physician may be liable to a medical malpractice claim for the exercise of that judgement.

---

[1]A defective product does not provide the safety which a person is entitled to expect, taking into account (a) the presentation of the product, (b) the use to which it would reasonably be expected to be put and (c) the time when the product was put into circulation.

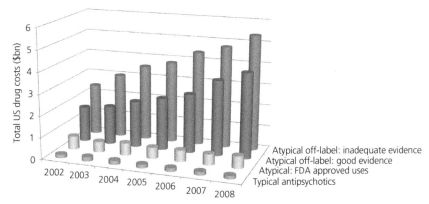

Figure 1.3 Costs associated with the prescribing of antipsychotic medications in the United States, 2002 through 2008, categorised by off-label status and level of supporting evidence. Reprinted with permission from Ref. [45]. © Nature Publishing Group/Macmillan.

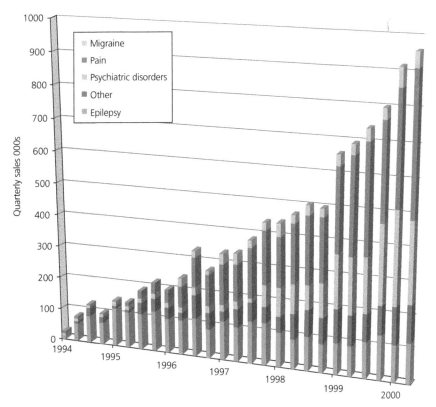

Figure 4.5 Neurontin (gabapentin) prescriptions, by indication. Graph drawn from data from Ref. [110].

Off-label Prescribing – Justifying Unapproved Medicine, First Edition. David Cavalla.
© 2015 John Wiley & Sons, Ltd. Published 2015 by John Wiley & Sons, Ltd.

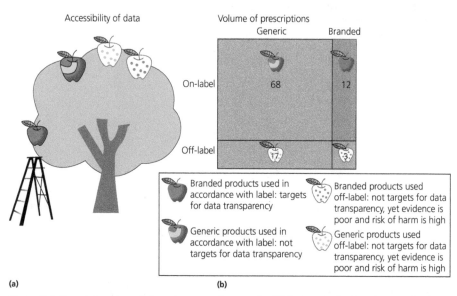

Figure 8.1 Generic and branded products, used on- and off-label, and whether they are likely targets for clinical data transparency **(a)** accessibility of data and **(b)** volume of prescriptions. The areas of the rectangle relate to numbers of prescriptions, normalised to 100%.

In *Richardson v. Miller*, 2000, a Tennessee woman and her husband brought a negligent malpractice action against a physician, after suffering a major heart attack during childbirth, resulting in permanent heart damage, consequent to the use of terbutaline sulphate. This drug is approved for the treatment of bronchial asthma; it was also used to delay labour by relaxing the uterine muscles, but the FDA had cautioned against this off-label use. At the appeal stage, the off-label nature of the use of terbutaline was considered on the basis that such information was essential in establishing the standard of care against which the defendant should be judged; a new trial was ordered, but there is no record of it actually taking place, suggesting that it was settled with agreed compensation package for damages before the new trial could begin. This is not surprising, given that the use of terbutaline for preterm labour is explicitly subject to FDA safety warnings.

In other countries with respect to negligence claims, the main issue similarly revolves around the evidence in support of the off-label use. In the United Kingdom, the courts generally do not hold unlicensed prescribing to be a breach of the duty of care, provided that treatment was supported by a respected body of medical opinion. The Bolam test which was formulated by a high court judge in 1957 has guided the courts as to the standard of care in medical negligence cases (*Bolam v. Friern Hospital Management Committee*). In addition, the more recent 1997 Bolitho case (*Bolitho v. City and Hackney Health Authority*) states that medical opinion should also be capable of withstanding logical analysis, which in this instance would imply that doctors should consider the risks and benefits of various treatment options, with regard to the evidence that is available and the nature of the clinical case.

The liability issues have been raised in certain quarters as reasons to be concerned at *insufficient* levels of off-label use, particularly in areas of high medical need, such as cancer. The UK peer and former advertising chief, Charles Saatchi, lost his first wife to ovarian cancer. His medical innovation bill sought to make it easier for a doctor to experiment in an effort to improve outcomes for cancer patients. As Saatchi puts it, 'Doctors deciding how to treat a particular case start with the knowledge that as soon as they move away from existing standards within the profession, there is an automatic and serious risk that they will be found guilty of negligence if the treatment is less successful than hoped'. His 2013 bill sought to differentiate in law between reckless experimentation and responsible scientific innovation. Ultimately, the Saatchi bill was unable to command government backing and failed to make headway in the House of Lords. However, it demonstrates that the acceptability of off-label medicine is related to the medical need of the patient and that in serious illness, patients are willing to tolerate greater medical freedom if it speeds up access to unregulated medicine.

Beyond ethics and legal liability, doctors prescribing off-label face another issue with respect to reimbursement, with payers sometimes declining to accept responsibility; it is even possible that the physician may be called upon to reimburse personally or be threatened to do so. Let's have a look at how this works in practice [40].

Reimbursement

Only rarely does a patient pay directly for their medicine. Normally, it is reimbursed by the insurance scheme (public or private) or by national health system. Occasionally, patients have to pay part of the cost, under a so-called 'copay' scheme, but it is generally only a minor proportion.

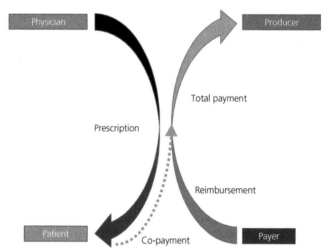

Figure 6.1 The flow of prescriptions from physician to patient is matched by a reverse flow of payment from payer to producer. The payer also determines the level of copayment by the patient, if any. The total payment is made up of the copayment and the reimbursement.

This is a complicated situation, but as money is at the centre of so much else, it is important to understand it at least in a general sense. Figure 6.1 shows the overall way in which the medicine is prescribed to the patient by the physician and the money then flows from the payer to the producer. The money for the reimbursement that the payer provides to the producer comes from general taxation or from insurance premiums: in other words, from the sick and the well alike. The total payment that goes to the producer is composed of the reimbursed component from the payer and the copayment from the patient.

Reimbursement is becoming a bigger and bigger issue as healthcare systems are becoming stretched by the twin costs of an ageing population and by increasingly expensive pharmaceuticals. A good example of this is the introduction in 2014 of the groundbreaking new treatment for hepatitis C virus from the biotech company Gilead, called Sovaldi™ (sofosbuvir), which is estimated to offer a cure for about 90% of patients but to cost $84000 in the United States for a 12-week course. Insurance companies have rallied to object to the cost of the product, which they say 'will break the country', since hepatitis C is a relatively common disease. In fact, they have calculated that if administered to all hepatitis C virus patients, this one drug alone would cost the US healthcare system $300 billion a year, more than the country now spends on all prescription medicines [218].

These high estimates have been questioned [219], with suggestions that the initial high costs will soon ameliorate because many patients acquired their disease before blood transfusions were screened for HCV and the current rate of new patients is far lower, around 45 000 per year. In addition, there are a number of other new treatments nearing approval, bringing the possibility of price-based competition into the equation. Nevertheless, regimes that were previously lax are finding strong reasons to look much more closely at what is reimbursed and what is not. One way of limiting spending on pharmaceuticals is to ensure that they are only reimbursed when their prescription is properly justified. When pharmaceuticals cost a few hundred dollars, pounds or euros a year, it is not so important whether the indication, patient or dose falls exactly within the regulatory label. The situation is entirely different when pharmaceutical treatment costs tens or hundreds of thousands of dollars a year. Increasingly, off-label prescriptions have become the target for more careful justification, and they are not automatically paid for under the rules and regulations of the payer. It may be at their discretion, or there may be specific restrictions on what can or cannot be reimbursed. Again, this varies country to country and in some countries depends on the particular insurance company.

In Germany, the statutory health insurance covers the costs of off-label therapies only in exceptional situations, such as for the most serious (but potentially curable) or very rare diseases. Exceptions also apply in cases where there is no alternative licensed option or if the local statutory health insurance steering committee has approved insurance coverage [217]. In France, there is a complicated system that offers different reimbursement conditions depending on whether it is prescribed in the community or hospital. Outside the hospital, off-label medication is not covered by insurance; in the hospital, cheap drugs are covered, but expensive ones are looked at much more carefully. In the case of oncology drugs, for instance, they must be included in the recommendations of the French national cancer institute (or exceptionally, based on published results and on local guidelines) [220]. In Italy, another complicated set of rules operates, with the national agency for drugs having established three lists (solid tumours, haematology and paediatrics) with accepted and covered off-label indications, to which are added individual cases which are covered in the hospital setting.

In Canada and the United Kingdom, the system is generally set up to reimburse off-label use. In Canada, outside Quebec, most provinces do not require a doctor to indicate the therapeutic diagnosis on the prescription. Unless reimbursement criteria are specified, most insurance plans do not even require that the pharmacist know why the drug is being prescribed, so that if patients are eligible for benefit coverage, the drug will most likely be paid for by their benefit plan. A similar situation to that in Canada applies for most prescriptions in the United Kingdom, which reimburses prescriptions that are on- and off-label alike. However, since the establishment of the UK National Institute for Health and Care Excellence (NICE), a much greater degree of control has been exercised on the reimbursement of drugs, so that their eligibility is constrained according to the indication and the patient; this is further dealt with later. In Japan, off-label drug use is not reimbursed.

Compendia

In the United States, off-label prescriptions are not routinely covered by private health insurance plans and managed-care organisations because they are regarded as experimental. However, the public healthcare system in the United States is more flexible. In 1993, the Omnibus Budget Reconciliation Act (OBRA) permitted the use of off-label anti-cancer drugs if provided with evidence of efficacy reported in the literature or in well-respected drug reference books. As a result, Medicare, which covers the medical costs for most patients with cancer in the United States, as well as Medicaid, will reimburse off-label drugs that are referenced in recognised compendia that aggregate published peer-reviewed original literature. Following this federal mandate, many states and other payers have subsequently adopted similar policies, and compendia are important components of the reimbursement policies of many healthcare providers in the United States.

The compendia function as gatekeepers; they are expected to be authoritative, comprehensive and timely sources of standards for the healthcare system and its reimbursement in the United States. However, an analysis of the off-label uses of a range of oncology drugs found quite poor results in the review process, with scanty evidence, inconsistency and irregular updates being significant problems [221].

Although compendia are theoretically a source of quality evidence for off-label use that should make value-based decisions easier for payers, in practice, they are a poor source of cogent evidence and have been criticised as contributing to poor decisions on coverage and reimbursement [222].

Oncology is a special case with high rates of off-label prescriptions for a number of reasons: first, there are many types of the disease, some of them extremely rare, making many rare cancers also orphan diseases for which there may be no preexisting licensed treatment. Second, the serious nature of the disease makes it particularly amenable to clinician experimentation, just in case a life could be saved. And third, there is a perception that all cancers share similar features, so that a drug that is effective in one is likely to also be effective in another. Bristol Myers Squibb's anti-cancer drug Taxol™, which qualified for orphan designation for AIDS-related Kaposi's sarcoma, was a substantial beneficiary of the OBRA legislation, since the largest market for Taxol is for breast cancer, a common disease that is definitely not designated as 'orphan'. However, anti-cancer agents that work in one type of cancer do not always work in another, and efficacy is not predictable: there are a number of important exceptions, as reported earlier with reference to bevacizumab (Avastin).

Other parts of the US healthcare system have also started to use compendia-based methods for qualifying reimbursement of off-label medicine. One of the component parts of the Medicare drug benefit system, known as Medicare part D, involves a certain requirement for patients to copay for their medication, with additional help for low-income groups. Medicare was originally intended to cover the over 65s, but under President Obama's Affordable Care Act, it will also cover under 65s with low incomes. It is growing substantially: by 2020, Medicare is projected to cost over $900 billion, with a similar cost attributable to Medicaid, the US healthcare support mechanism for patients with low incomes. Under the Medicare Part D benefit, off-label prescription costs are not reimbursed to the patient unless they are included in one of the compendia. There has therefore been a dramatically increased importance attached to the legitimisation of off-label treatments by inclusion in the compendia.

Pharmaceutical companies also recognise that inclusion in a compendium is an easier way to gain reimbursement coverage than regulatory approval and strive to establish good relationships with compendia staff. It works similarly to the way that journalists' work load can be alleviated by public relations firms providing pre-packaged pieces about their companies. Pharmacists who write compendia are usually quite receptive to receive well-organised packets of scientific articles, abstracts and other information supportive of a particular drug's off-label utility. Using company-provided articles saves time, but if assertions contained within the information are not subjected to a rigorous review, there is a great danger that the conclusions about the validity of a certain off-label use will both be false and become embedded without sufficient independent review. The fact that the gatekeepers to this situation can be relatively junior individuals rather than senior panels of experts is a particular element of concern. This situation sits right on the blurred line between pharmaceutical marketing and pharmaceutical fact. Indeed, some activist groups have specifically highlighted this issue as good for corporate profits but poor for patient care [140].

Outside the government healthcare arena, private insurance companies have their own difficulties and inconsistencies with deciding what is reimbursable and what is not. Let's take the example of eculizumab (Soliris™), a monoclonal antibody from the biotech company Alexion for the C5 protein, which down-regulates specific immune reactions. It is approved for the ultra-rare and life-threatening diseases of paroxysmal nocturnal haemoglobinuria and atypical haemolytic-uraemic syndrome but has the dubious honour of being the most expensive pharmaceutical in the world, at over $400 000 per patient per year. With the label restricted to these two orphan diseases, there are only a few thousand patients worldwide. Needless to say, there is a long list of potential additional uses, all also rare and under the umbrella of being off-label use, but with varying degrees of evidence in their support. Among these conditions, only a short list of diseases is covered by insurance, but the list is different depending on the insurer. So, dermatomyositis is an off-label use covered by Blue Cross and Blue Shield of Florida but considered investigational and not covered by Aetna; the same goes for idiopathic membranous glomerular nephropathy. However, for the patient that can afford it and is prepared to pay, the long list of conditions is available for intervention. As time develops, one can expect the list of reimbursable off-label conditions to be extended. Alexion is one of the world's most commercially successful pharmaceutical companies, with only just over 1000 employees and a market capitalisation of just over $27 billion (as of 20 January 2014).

NICE

In the United Kingdom prior to 1999, the licensure of a drug by the regulatory authorities meant that it was effectively reimbursed for all uses. Most prescriptions in the United Kingdom do not carry the prescribed indication, and the pharmacist is effectively obliged to fill it without question. However, in 1999, the UK government established a new body called the NICE. Its role was to examine the cost-effectiveness of new treatments for the UK NHS. Faced with increasing costs for new pharmaceuticals (particularly new biological drugs) that were far beyond old-style generics, the UK government wanted to establish whether the marginal improvements in clinical outcome warranted the large increases in price. NICE's evaluation centres on the additional quality-adjusted life years (QALYs) that new drugs offer, and it is generally believed that it will only approve for reimbursement drugs that offer QALYs costing below a ceiling that falls between £20 000 and £30 000. Recently, however, the organisation indicated that other benefits to society would be included such as factoring in whether a drug will help a patient return to work, at which point they would cease to be a burden on the state and return to paying taxes. There is evidence that NICE guidance is sometimes referred to by countries other than the United Kingdom, because of the robust methodology perceived to inform their review process and because their guidance can be available before a drug has gained approval in such other countries.

At the time of its establishment, there was public outcry from two quarters. First, the pharmaceutical industry complained that this was an additional hurdle to clinical use of their products: beyond regulatory clearance, in itself an increasingly challenging benchmark, the effective market would remain tiny until it became accepted for NHS reimbursement. This reflects the very small size of the private healthcare sector in the United Kingdom. The second group opposed to the change was, also unsurprisingly, patients. Their view, though obviously partisan, was also supported by the head of the EMA, who raised the possibility that organisations such as NICE were impeding patient access and threatening the European single market. Many felt the additional delay would deny them drugs that had been proven to be safe and effective purely on cost grounds. This was most acutely felt for serious diseases, and in some areas such as breast cancer, significant patient advocacy groups set about trying to pressure NICE into reassessing their position, in this case on the drug Herceptin™ (trastuzumab). For the sake of completeness, NICE's position was not that the product was not safe and effective, but just that it did not prolong life sufficiently (in many cases by only a few months) to justify the cost. But for many patients without other treatment options, the calculation of how much that few months was worth was an inhuman way to deal with human existence itself.

Like so many of the regulatory barriers that affect the pharmaceutical industry, the genius of commerce has found ways to work with, or around, the new requirements. In fact, the approval by NICE for Roche's Herceptin was partly orchestrated by corporate support for the very patient advocacy group composed of breast cancer sufferers who stood to benefit. But the most apposite study in how the industry used NICE to its own advantage is told through the history of Lucentis; it also turns out to involve off-label uses, not for the product itself, but for a cheaper competitor.

Lucentis is the trade name for another monoclonal antibody, ranibizumab, this time from Genentech but marketed in the United Kingdom by another Swiss company, Novartis.[2] It is approved for the treatment of wet AMD, a progressive retinal disease which affects millions and is a leading cause of visual loss in persons at least 60 years of age; it is also approved for other eye diseases including macular oedema. Lucentis is very (inordinately?) expensive, costing just under £9000 per eye per year, and because of its price, it had a difficult time being approved through NICE. Initially rejected, Lucentis was eventually accepted by NICE for reimbursement in 2008 after pressure from patients, but also reflecting the substantial clinical benefits with the drug.

Lucentis is a monoclonal antibody for the vascular endothelial growth factor (VEGF) receptor, a property shared with the anti-cancer drug Avastin (bevacizumab). VEGF is a signal protein produced by cells that stimulates blood vessel growth, known as angiogenesis. Wet AMD involves abnormal angiogenesis in the back of the eye, which leaks blood and fluid into the retina, causing distortion of vision that makes straight lines look wavy. Angiogenesis in the context of cancer involves the solid tumour developing its own blood supply to assist the tumour's growth. Both Lucentis and Avastin block the VEGF receptor: Avastin controls blood vessel growth in the solid tumour, in, for instance, the colon, whereas Lucentis does the same thing in the back of the eye.

Despite their biochemical similarity, the two drugs differ substantially in price, with Lucentis being the more expensive product by a factor of between 10 and 15. Avastin is approved for the treatment of certain types of colon, lung, kidney and brain cancer, but not for any ophthalmic use. In addition, Lucentis is required to be injected directly into the eye, a route of administration that is not approved for Avastin. Nevertheless, the price differential alongside mechanistic similarity led many clinicians (not just in the United Kingdom) to consider the use of Avastin for wet AMD. To do this required dividing up the high doses of Avastin normally given to cancer patients into much smaller doses appropriate for direct injection into the eye. A strictly sterile environment is required for this process, which precludes it being safely done outside specialised manufacturing suites (this hasn't prevented various amateurish attempts in what have turned out to be non-sterile environments, with disastrous consequences, as discussed on p. 112). After some considerable (but poorly documented) experience with the anti-cancer drug for this eye condition in the United States, the NIH decided, at their cost and in the interests of public health, to evaluate the two drugs side by side.

The NIH trial which compared Lucentis to Avastin in over 1100 patients posted headline results of similar efficacy. The trial also showed that a reduced dosing frequency of either drug provided similar levels of improvement, potentially producing a further fivefold reduction in drug use (and cost). At this point, the regulatory

---

[2]To complicate things even more, Genentech was taken over in 2011 by Roche, the manufacturers of Avastin.

barriers imposed by NICE were a significant reason why wet AMD patients in the United Kingdom were not generally able to receive Avastin, since although the drug was approved by the UK regulators (the MHRA), it had been rejected by the NICE as being 'too expensive' for any of the other oncological uses.[3] This posed a regulatory dilemma for the UK government, since the body established to promote cost-effectiveness in the United Kingdom looked like being the same body that was standing in its way of a cheaper alternative to Lucentis. A significant problem with the approval of Avastin for eye disease was that it had never been approved for any such indication by the UK regulatory bodies which normally precede NICE evaluation. Despite its lack of passage through a formal regulatory review process, certain local NHS authorities proposed a new commissioning policy to fund Avastin for wet AMD, and there was a real prospect of national policy disarray; Novartis responded by the initiation of a judicial review. It therefore fell to the Department of Health (DoH) to ask NICE to look at this as an exceptional circumstance.

Meanwhile, negotiations on price between Novartis and the NHS continued. Novartis, while admitting the therapeutic equivalence in efficacy terms between the two agents, maintained the NIH trial results revealed a 30% higher risk of adverse events with Avastin compared to Lucentis. These, apparently, were due to the longer systemic half-life for Avastin compared to Lucentis, giving rise to serious reactions such as arterio-thrombotic events, systemic haemorrhage, congestive heart failure, venous thrombotic events, hypertension and vascular death. The risk of these events is nevertheless low, but pointing to safety as an issue had more than a public relations impact.

When finally NICE did respond to the DoH request, it said that there would need to be a regulatory analysis of safety for Avastin in wet AMD to be conducted before it could undertake its cost-effectiveness review. In essence, it was saying exactly what, in part, this book says: that the approval of Avastin for the treatment of cancer does not guarantee the safety in other diseases. As of now, Avastin remains unapproved in the United Kingdom for use or reimbursement for wet AMD. Novartis had found a way to use this newly established body, initially characterised by the industry as a death knell for commercial pharmaceutical innovation, as a means to stifle cost-effective treatments – rather than do what it was set up to do, exactly the opposite. The situation elsewhere in the EU is still evolving. The French and Italian governments have specifically crafted legislation to favour the unapproved use of Avastin as a cheaper alternative to Lucentis for the treatment of AMD [223]. The European Federation of Pharmaceutical Industries and Associations (EFPIA) has responded with a position paper vehemently opposing this move, pointing out the destabilising consequences for the European regulatory framework, and its potential to compromise patient safety and create legal uncertainty. The legal uncertainty is described in more detail at the end of this chapter (in Section "Cost as a driver for off-label medicine") but briefly, it involves a European legal judgment that allowed unapproved drugs only when completely essential and based on therapeutic need, not financial considerations. It is quite possible therefore that the French and Italian moves will face legal challenge [224]. At a wider level, the machinations of these governments in the role of payers may also undermine attempts by the very same governments as regulators to police off-label marketing of medicines; indeed, a whiff of hypocrisy is detectable.

---

[3]There were separate means, through a special Cancer Drugs Fund or specific Individual Funding Requests, by which patients could obtain it, and even an intravitreal formulation available through the London Moorfields Hospital for specific patients, particularly those with eye conditions outside AMD for which Lucentis was not licensed and Avastin offered a potential solution.

The other point about NICE, which I touched on earlier, is that under the Therapeutic Appraisal scheme, it establishes an indication-specific set of reimbursement rules for the NHS, which previously had been absent from the UK scene. In practice, this applies to all newly approved pharmaceuticals, so that off-label use of a new drug is not generally reimbursed. However, NICE also provides broader guidance around therapeutic areas through its Clinical Guidelines programme, and sometimes that guidance includes off-label options. For instance, it includes options for paediatric care even if the regulatory label does not include children. But in all cases, it does look intensively at the available information so that all decisions are evidence based, and in many cases, paediatric and adult care options are different. NICE published its first off-label assessment summary in 2012, because of 'a lack of nationally available, good quality information about using unlicensed and off-label medicines'. This is a very welcome new addition to the activities of the organisation, and for the first time, we have an authoritative voice enabling us to differentiate evidence-based off-label uses from the snake oil.

Compassionate access

Between the first clinical trial and the approval of a drug, there is a long period of uncertainty, but for a person with a life-threatening disease, these developmental drugs may either offer a cure or curtail the disease process long enough for a cure to be found. Ever since HIV/AIDS in the 1980s, patient advocacy groups have pressed for earlier access to developmental drugs on the basis that while the science goes on, people die. As a result of this pressure, in the United States, there is a compassionate use route for developmental drugs to be used for treatment, separate from the clinical trial process, which is primarily for testing purposes. The first compassionate access was granted for a drug called AZT, originally created to treat leukaemia, but discontinued and subsequently developed to become the first FDA-approved treatment for HIV/AIDS. During its formal development, almost one-third of AIDS patients in the United States were allowed access to AZT for free.

Under the subsequently established 'treatment IND' (investigational new drug) programme, the FDA may approve use of an investigational drug by patients that are not taking part in clinical trials. Such drugs can be given for the treatment of 'serious or immediately life-threatening disease[s]' if there exists 'no comparable or satisfactory alternative drug or other therapy'. There are a couple of additional requirements related to the active development of the drug and, most importantly, the agreement of the company to supply the product.

Not all companies participate in compassionate use programmes, especially relatively early in drug development. This is primarily because they wish to retain control over the way in which the compounds are developed and wish to avoid serious adverse events which could derail their clinical trials. Patients view policies like this as heartless, yet the use of a developmental drug is not always a one-way bet. Consider the position of a manager at Novo Nordisk, developing their factor VIIa drug for a rare form of haemophilia, who receives a request for a sample of the drug before it has been approved for any use, in order to address a different form of haemophilia. In some ways, it might seem a reasonable request, except as we now know from the history of factor VIIa that off-label uses can cause excess strokes and heart attacks and are ineffective. This kind of pollution of the evidence base *before* approval can be fatal for a drug development programme. A different situation exists *after* approval, when drug companies are happy to sell the product without questioning what it might be used for.

Another argument against compassionate use is that it is poor at capturing outcomes, the data from which inform future generations of patients on a wider scale.

If people simply have access to untested medicines without those trials, we will never learn which ones are effective and how best to use them. In addition, in diseases where it is difficult to recruit into clinical trials, compassionate access programmes effectively compete for the same patients needed for the regulatory trials. So, a balance needs to be struck.

Nevertheless, the compassionate use programme has a worthy aim, and it generally works well on a limited scale. It is not the same as off-label use since the products are not offered for sale, but given for free (another reason drug companies are not so keen on them!).[4] Occasionally, it offers extended lives for people with terminal illnesses. Initiatives like the one in operation through the FDA have been welcomed and copied, so that there is a similar programme in place in Europe and alternately named 'Special Access' programmes in Canada and Australia.

Cost, as a driver for off-label medicine

From the previous section, the cost of medicine is a factor in how and whether it is reimbursed, and can also influence the choice of drug. But in this regard, the payer's and patient's interests are not necessarily aligned, even though, of course, resources for providing healthcare are taken from patients' taxes or insurance premiums. The patient is generally reluctant to think that their health should be constrained by cost of treatment, since they are not usually paying for it directly. They tend to align themselves with the interests of the breast cancer patient advocacy groups arguing for access to Herceptin, set against a 'miserly' attitude of the UK NHS. On the other hand, governmental and insurance payers are acutely aware of the burgeoning pharmaceutical bill. In addition to the separate interests of the patient and payer, then there is the person in the middle making the prescribing choice.

Cost and value in pharmaceuticals are completely different things, and this is no more obvious than in the case where a single drug has two different applications. Clearly, the value to health of a pill to prevent cancer is orders of magnitude more than a pill to treat toothache. But if it is initially introduced and priced for the lower-value application, it becomes much more difficult to justify an investment to develop it for the secondary indication. An additional issue arises when generic drugs have two or more potential uses. These issues are clearly displayed in the remarkable versatility of aspirin, as a drug to treat inflammatory pain, to prevent strokes and to prevent cancer. The product is of course widely available over the counter, and the same product (i.e. the very same tablets) can be used for all therapeutic endpoints. The development of the drug for the anti-thrombotic, and now for the anti-cancer use, has relied on public funds, since it makes no commercial sense to spend many millions to prove it works for a new use when a competitor company could sell the same product without spending this money. In consequence, the development of the secondary and tertiary uses of aspirin has taken decades. The availability of patents and restrictions on normal competition may offend our sensibilities of equal access to healthcare for all, at reasonable prices, but they also facilitate faster development cycles, albeit with a cost attached. Alliteratively, when patent-protected products are not possible, patients need patience.

This problem affects the acquisition of clinical trial information on many potential new uses of existing drugs (such as those listed in the DrugRepurposing.info database

---

[4]Drug companies normally impose 'administrative costs' on any order, which can be significant.

available at http://drugrepurposing.info). It is a great pity, since these products could enter into clinical use with much less developmental spend than new drugs, which have to undergo the full range of discovery and toxicological tests before embarking on a risky trial process. Pharmaceutical companies often opt to pursue the discovery and development of an entirely new agent purely because of the commercial protection and return it offers. Absent the commercial driver, there is no incentive to invest in even the shortened adaptive process to prove an existing drug has utility beyond its initial use.

There might be a middle way, however, in certain circumstances. One possible route is if the secondary development is supported by charitable groups, which aim to identify new treatments for specific diseases. The Michael J Fox Foundation (a US charity funded by the eponymous actor who has developed early-onset Parkinson's disease) supports drug development for Parkinson's disease based on this strategy; another programme has recently begun funded by the UK Alzheimer's Disease Society. There are other 'disease agnostic' organisations such as Cures Within Reach, which is actively engaging in the development of secondary uses for existing drugs. Charitable organisations such as these are very clearly both funded by and operated on behalf of patients. It will be interesting to see how these programmes pan out. One problem could be that even if initially successful, what might happen if the charitable funds run out before full development and regulatory approval can be gleaned? Normally, mid-stage products of this kind can be licensed onto a large company for later-stage development and commercialisation. But absent a patent and a commercial differentiating factor, incompletely developed products may still languish without a marketing partner. These developments either need to go the whole way or their backers think carefully about how they can attract another organisation to take the products through the regulatory process. Cures Within Reach has actually considered this problem and arrived at a proposed solution, which I will mention in Chapter 5 (see p. 168).

A second area where cost is an imperfect driver for improved healthcare is in motivating physicians to choose the drugs on the basis of cost rather than medical advantage. When the UK GMC were renewing their prescribing guidelines in 2010, they specifically proposed to relax the rules to allow off-label prescriptions 'where, on the basis of authoritative clinical guidelines, a doctor judged that the medicine was as safe and as effective as an appropriately licensed alternative'. The change would have allowed the possibility of significant cost savings, particularly for rare diseases where a doctor might be faced with the alternatives of a specific high-cost treatment that had been approved for the condition, alongside a much cheaper off-label generic. An example of where this situation might arise is provided by the example of hydroxy-carbamide. This drug is available as a generic medicine licensed for chronic myeloid leukaemia and cancer of the cervix but as a substantially more expensive product, Siklos™, for the orphan indication of sickle cell anaemia, the regulatory license and orphan designation for which is held by the UK company Norton.

The proposal was sent out for consultation and met with divergent responses. Though generally supported by many doctors, there was substantial opposition from both the UK pharmaceutical company lobby group and from UK regulators. As a result of these comments, the GMC sought legal advice in relation to the EU Directive on medicinal products for human use. The EU has specifically looked at whether cost should be a determining factor in choice of prescription medicine.

In March 2012, the Court of Justice of the European Union ruled that a Polish law, allowing the import and sale of unapproved and less costly medications to similar, approved drugs under the 'special need' exemption, was a breach of EU law. The ruling noted that the exemption was only to be used when completely essential and based solely on therapeutic need. If licensed alternatives are available, there cannot be a requirement for 'special needs', and financial considerations cannot be used as a justification. As a result of this legal analysis, the GMC's proposed changes to facilitate generic substitution were abandoned, and the guidelines referred to earlier in this book (see p. 38) reverted to the previous position, which stipulated that off-label prescriptions should not be used when there was a licensed alternative. This legal judgment might also be referred to in the continuing controversy around Avastin and Lucentis described above, since despite the substantial price of the latter, it has been subjected to the rigours of formal regulatory clearance, and its substitution by the cheaper product cannot be justified on therapeutic need alone.

The third point in regard to cost that needs to be borne in mind is the lack of joined-up thinking with regard to overall cost. We have seen in the previous chapter how serious adverse events are more common with off-label drugs, yet the overall burden from a pharmacoeconomic perspective is rarely included in the debate over whether to use off-label generic alternatives to licensed medicines. The consequence may give rise both to the possibility of additional costs associated with ameliorating the adverse event (such as additional time spent in hospital), as well as the costs of poorer therapeutic outcomes.

When the EMA looked at this issue, they noted with some concern the increased frequency of adverse drug reactions in the paediatric population and called for greater pharmacovigilance to address the problem which they attributed to off-label or unlicensed medicines in children [118]. We can appreciate the true pharmacoeconomic cost of the wrong prescribing decision resulting in a chronic or lifetime deficit when lawsuits seeking tens of millions in damages are brought to bear (like the man who was blinded by the non-sterile injection of Avastin into his eye on p. 112). But on a more mundane level, even the possibility of an additional night or two in an intensive care unit in hospital makes acceptable the additional cost of a drug specifically developed for a certain condition, relative to an off-label generic with much more uncertain risks and benefits.

I will return to the issue of cost and pricing in the last chapter, as I seek to find ways to attract commercial, as well as non-commercial development of secondary uses of existing drugs. But before doing so, we now move on to consider how pharmaceuticals are regulated, specifically in the context of off-label uses.

CHAPTER 7

The role of regulation in off-label medicine

Regulatory approval can involve dozens of scientists poring over extensive databases of studies in animals, toxicologic evaluations and clinical trials [225]. But regulatory involvement does not stop after marketing approval has been authorised. In a variety of ways, the authorities continue to assess post-marketing studies, monitor safety and keep an eye on the marketing activities of the companies to whom approval has been granted. In addition to the primary approval to put a drug on the market, changes to the label can occur as a result of a supplementary new drug application for a secondary use. This process does not just allow the new use to be promoted, but it also ensures reimbursement from third-party payers, helps to get that use included in formularies and gives the medical community more complete information about the product.

In the United States, the FDA is empowered by law to review drug products for safety and efficacy. FDA approval has been required since 1938, at which time manufacturers were obliged to include the complete list of ingredients in the label along with instructions for safe use. Concern about manufacturers making accurate declarations of their ingredients and the uses for their products came from a much earlier episode, in which it was discovered that Mrs Winslow's Soothing Syrup for teething and colicky babies was actually composed of morphine, and its overuse led to the deaths of many infants. FDA regulations were further enhanced in 1962 with the additional requirement that new drugs be reviewed for both efficacy and safety, rather than safety alone. As referred to in Chapter 1, one of the primary motivations behind the extra regulatory requirements was the thalidomide tragedy that unfolded in the immediately preceding years. It was not until 1976 that the FDA had the authority to regulate medical devices.

The general regulatory scheme in the developed world for pharmaceuticals has become tighter as time has passed. In order to get a licence, manufacturers nowadays must submit whole rafts of information on the results of clinical trials, adverse reactions, manufacturing and quality control processes, and information about how the medication will be marketed. Then, medicines are divided into those which can be supplied without a prescription, either for general sale over the counter or with the advice of a pharmacist, and the remainder which require a prescription. This latter group is reserved for the medications carrying the greatest degree of danger, require intravenous injection, as well as those which are most likely to be administered incorrectly, or are addictive and so on. Prescription-only medicines are ineligible to be directly advertised to the public (except, as we have seen, in New Zealand and the United States).

The product label which accompanies every prescription drug is the result of negotiations between regulatory authorities and manufacturers, and includes a summary of the safety and effectiveness for specific indicated clinical conditions, including

Off-label Prescribing – Justifying Unapproved Medicine, First Edition. David Cavalla.
© 2015 John Wiley & Sons, Ltd. Published 2015 by John Wiley & Sons, Ltd.

dosing, duration of administration, special safety and/or effectiveness considerations, drug interactions, data from paediatric or geriatric experience, and specific warnings or contraindications. The complexity of the label has grown enormously with the years, so that by 2006 the average number of potential adverse reactions listed in a given product label was nearly 70 per drug. There has been further growth since 2006 of around ~5–7% annually, as a result of the increase in information derived from post-marketing surveillance and other regulatory mandates, such information being provided by manufacturers.

There are also post-marketing processes for the identification of adverse drug reactions, which in the United Kingdom is a requirement for a minimum of the first 2 years after the grant of a licence and subsequently becomes voluntary. This is the so-called 'Black Triangle' scheme, which is in the process of being extended across the entire EU. Marketing authorisations are time limited, and manufacturers must apply for a renewal of the product licence at least every 5 years. And, of course, the licence is for a drug to be used in a specific context, for a certain type of patient (normally an adult aged 18–65), for the treatment of a certain condition, for a certain time and at a certain dose. Given the high degree of control exercised by the regulatory authorities in granting the licence, it is somewhat surprising that the use of the drug, the age of a patient, the length of treatment or even the dose can be actually different in practice from that which the licence stipulates. It makes you wonder what was the point of the regulatory process in the first place.

Regulators do not regulate medical practice

When it was established, and in deference to the medical profession, the US Congress was nervous about ensuring that the FDA would not regulate medical practice and managed to insert a clause in the 1938 bill to that effect (under 21 US CODE § 396). But even in Europe, where the same nervousness did not prevail, the role for the EMA has also not overreached into the day-to-day practice of the medical professional (the restrictions such as they exist come from the professional guidelines, not from governmental bodies). So, while the regulatory authorities are very strict about the manufacturing standards and documentation supporting the medicines we take, they do not regulate the way they are prescribed.

The FDA does say [226]:

> If physicians use a product for an indication not in the approved labeling, they have the responsibility to be well informed about the product, to base its use on firm scientific rationale and on sound medical evidence, and to maintain records of the product's use and effects.

But there are no powers to enforce this 'responsibility'.

This situation is rather like having a very strict regime for the manufacture of cars, with very precise specifications for the materials used, the processes employed and so on, but having nothing in place for the way they are driven. Public health clearly has an interest in both the way that cars are driven, and also in the way that medicine is practised. But the regulatory law nevertheless is constituted to leave US physicians alone. Some legal commentators have suggested that the exclusion of the FDA's powers from the practice of medicine makes the issue of whether a treatment is off-label, unlicensed or on-label irrelevant for a physician [215]. I suppose, to take the automobile analogy a little further, this could be akin to the car driver being allowed to drive without seat belts, even if fitted.

This is actually an interesting analogy, since it involves a public debate that has changed direction within the last 30 years: in the 1970s, various unsuccessful attempts had been made to legislate in the United Kingdom for compulsory wearing of seat belts (when fitted). That changed when, in 1983, seat belts became compulsory for drivers, and later for all passengers in a car. Society's demands of a safe environment for drivers had outrun the principle of individual freedom. Similar debates with similar outcomes have now taken place around most countries in Europe and North America. When considered in this way, society's attitudes around off-label medicine, and their silent acceptance thereof, may also be on the point of change. Statements suggesting that the regulatory status of a medicine is irrelevant to a doctor might now appear to be outdated legalistic sophistry; physicians obviously do have an interest in knowing the regulatory status of the prescriptions they write (and so do their patients) although there is no automatic legal penalty (yet) for failing to know this information.[1] I will deal with how false perceptions like this should be tackled in more detail in Chapter 8.

There is a growing realisation by regulators of the problem associated with the off-label use of drugs, but they are powerless to intervene in the doctor–patient relationship. When, in 1972, the FDA proposed extensive regulation of off-label prescribing, there were vehement objections from the medical profession, and the proposals were 'kicked into the long grass'. They did not emerge again until the early 1990s, when the agency made an attempt to reconsider the 1972 proposals. The American Medical Association was again strongly critical, and the agency commissioner at the time, David Kessler, felt vilified by his detractors: the initiative was again dropped.

Despite the legal disempowerment of the US regulatory authorities to interfere in medical practice, the same obstacle does not work in reverse. In other words, medical practice can sometimes drive regulatory authorisation. To some extent, this may not surprise you; however, you would expect the normal regulatory standards to apply, so that consequent upon the development of medical practice around a particular off-label use practised by physicians, large-scale clinical trials would need to be undertaken and then a rigorous safety and efficacy assessment made. While this course of events does take place, there is also a particular example where the US regulatory authorities were persuaded to operate against normal protocols, in order to adapt to medical practice, practice that turned out to be pretty unsafe.

The case involves the use of amiodarone for cardiac arrhythmias. Amiodarone was initially developed in 1961 by the small Belgian company Labaz. Its effects in arrhythmia were discovered in Oxford by Bramah Singh but taken up by the Argentinian physician Mauricio Rosenbaum and then by physicians in the United States, who began prescribing amiodarone in the late 1970s to their patients with potentially life-threatening arrhythmias [228]. By 1980, amiodarone was commonly prescribed throughout Europe for this purpose, but in the United States, it remained unapproved, since reports had begun to emerge of a bizarre series of side effects – including difficult thyroid disorders, skin discolouration and potentially life-threatening lung toxicity. At this time, US physicians had to obtain the drug from

---

[1]Interestingly, in March 2014, the House of Representatives of the US State of Oklahoma voted overwhelmingly in favour of a bill to ban the off-label use of drugs to induce abortions. Oklahoma is among five states – the others are Arizona, North Dakota, Ohio and Texas – that have sought to restrict medical abortions by limiting or banning off-label uses of drugs. In arguing for the passage of the bill, proponents argued that the law would reduce patient harm, since a number of women had died while being prescribed drugs to induce abortions [227].

Europe or Canada on a 'compassionate use' basis, which didn't involve any payment, but its growing popularity meant the manufacturer was becoming increasingly disadvantaged. In the mid-1980s, there was twin pressure on the FDA, both by the European manufacturer and by the medical establishment, as a result of which in December 1985, amiodarone was finally approved for 'life-threatening arrhythmias for which no other treatment was feasible'. Amiodarone thereby became one of the very few drugs approved by the FDA in modern times without rigorous randomised clinical trials.

After its approval, like so many other drugs, amiodarone became widely used off-label, mainly for atrial fibrillation rather than the much more serious ventricular arrhythmia for which it was intended. This is tragic since there are safer alternatives for the former condition; recent estimates have put the extent of this off-label use of amiodarone at around 80% (i.e. the proportion of its use in accordance with the regulatory label is only 20%). Now, the toxicities which had been discovered before its approval came back to haunt the product. In particular, it has been associated with the development of conditions like pulmonary fibrosis (even after cessation of the drug), a serious disease which carries a prognosis of death within 5 years of diagnosis. The FDA has issued warnings about the pulmonary toxicity, and other side effects associated with amiodarone use, which stipulated it should only be used for 'life-threatening arrhythmias for which no other treatment was feasible'. It has required a patient information leaflet to be provided at the time of prescription, but physicians continue to use the product off-label; patients therefore also continue to develop serious toxicities, lawyers seek to acquire clients around which to frame class action lawsuits, and patients continue to look to the courts to compensate them for the harm they have suffered. Sometimes, the compensation is sought by the patients' families because the patients themselves have died.

Faced with limitations around their ability to control off-label use, but conscious that patients are being adversely affected, the FDA and EMA have sought other ways to exert regulatory pressure. These means fall into two groups: first is by controlling the way in which the pharmaceutical companies ('marketing approval holders') may influence prescribers, and second in collecting information from patients about their experiences, particularly insofar as safety is concerned, and using this information to warn prescribers about hazards ('pharmacovigilance'). I shall reserve a full discussion of the latter until the final Chapter, but it is now time to look in detail at the issue of regulatory policing of off-label marketing.

Off-label marketing

Before taking you on a trip to reveal some quite unpalatable activities at various large pharmaceutical companies, I want to point out a logical inconsistency. What we have is an allowance within our society, and indeed a very strong belief by practitioners, that off-label medicine should not be regulated while at the same time, a near-universal belief (by both patients and practitioners alike) that 'on-label' medicine should be regulated. In essence, we have a requirement that new drugs are assessed for safety and efficacy, while new uses of old drugs (or the same uses in different patient populations) are assumed to be safe and not required to pass any efficacy test (or only very limited ones) before being prescribed. If that is all acceptable for us, why is it unacceptable for companies to promote drugs in ways that accord with the way practitioners may use them? I will leave that question hanging for the moment, but I will refer to this incongruity later in discussing a legal challenge to the attempts by the regulatory activities to control off-label marketing.

The reason why the regulatory agencies are motivated to prevent the marketing approval holders from engaging in off-label marketing is that they are concerned that such behaviour can lead to widespread uses of drugs without a sufficient evidentiary basis for their safety and efficacy, and thereby expose patients to uncertain benefits and the prospect of adverse effects. Lest there be any doubt about it, pharmaceutical marketing does influence prescribing behaviour, as a recent review of the effect of changes to promotion of anti-depressants on prescription volumes shows [229]. Parenthetically, it would be somewhat absurd for companies to spend huge amounts on marketing activities if they did not firmly believe that this expenditure could improve sales. However, in addition, regulators are concerned that the freedom to engage in such marketing activities could undermine the incentives for manufacturers to conduct clinical trials necessary to achieve wider regulatory approval for their products.

There seem to have been divergent responses on both sides of the Atlantic to the problem of off-label marketing, with by far the most muscular sanctions to be found in the United States. Before the 1980s, the FDA imposed relatively few restrictions on companies' off-label marketing activities, but since then, the rules have tightened considerably. During the 1980s, public and congressional concern about these practices grew but faced with staunch opposition from the medical community (in particular the AMA) to government control of physicians' freedom to prescribe, any controls needed to be focussed on how medicines were supplied rather than how they were consumed.

US Congressional hearings in 1990 led to a change of the law so that manufacturers were prohibited from discussing off-label uses with providers and could not distribute written materials that mentioned an off-label use. In 1997, this was soon changed again, with the Food and Drug Modernization Act, so that a manufacturer could 'disseminate to a health care practitioner…written information concerning the safety, effectiveness, or benefit of a use not described in the approved labelling of a drug or device'; but the company could only do so if the off-label use described therein was included in a filed, or soon-to-be-filed, supplemental new drug application [230]. There were two other routes by which this information could be disseminated. One was if the physician asked for it from the pharmaceutical company; in addition, he or she could obtain it from a third party. Manufacturers could disseminate off-label use information to a pharmacy benefit manager, health insurance issuer, group health plan or federal and state government agency; but they were prevented from distributing off-label use information to patients or the general public.

In 2009, the law changed again, and the FDA issued a new guidance document that tightened the earlier rules. Companies were now allowed to distribute peer-reviewed scientific articles and texts describing off-label uses, subject to several conditions. In particular, the aim to file a supplementary regulatory approval was removed from the 1997 guidance. There is an interesting corollary to this change, which relates to the story of Aspreva, one of the most successful of biotech companies for founders and investors alike: put succinctly, Aspreva could only have happened between 1997 and 2009, since its business model revolved around the ability of a company to promote an off-label use, provided it was actively seeking regulatory approval for such use or proposing to do so soon (see Box 7.1).

The FDA is charged with monitoring pharmaceutical marketing practices, and the Department of Justice with enforcing the law, which is composed of three parts. First, off-label promotion violates the Food, Drug, and Cosmetic Act (FDCA) since the product is considered unapproved for the new indication and also a 'misbranded' medicine. Second, off-label promotion is considered healthcare fraud under the False

---

BOX 7.1

Aspreva was a West Coast Canadian company founded in 2003 by a three-man team composed of a scientist, a marketeer and a businessman. It was established with the idea of developing the transplant rejection drug, mycophenolate mofetil, for the orphan autoimmune disease lupus nephritis (and a couple of other even rarer diseases). Soon after it was established, some academics published clinical results showing favourable effects in a small trial. Aspreva then approached Roche, who marketed mycophenolate mofetil under the brand name CellCept. At the time, if a company was pursuing development of a product towards regulatory submission for a new use, the FDA allowed it to market the drug for that use; as we saw, these rules changed in 2009. By engaging with Aspreva, Roche then had the cover it needed to promote CellCept™ for lupus and the other two orphan diseases. Aspreva benefitted enormously from this deal, which they signed with Roche in 2004, since it allowed them royalties on sales of CellCept™ outside the primary indication of transplant rejection of 50%. As a result, Aspreva became a meteoric success. They became a stock market darling: from their initial public offering in 2005, when they were valued at $484 million, they drew in royalties of over $200 million in 2006 and continued to increase in value until sold to the Swiss company Vifor in 2008 for around $900 million. The founders were said to have walked away with over $100 million each.

Claims Act at the point when false reimbursement claims are made, either to the US government or to an insurance company. And third, the Federal Anti-kickback Law prohibits corrupt practices in relation to inducements or bribes of physicians by companies.

Off-label fines

The US enforcement of the law against pharmaceutical companies in relation to off-label activities has been vigorous in the past 10–12 years, and billions of dollars have been collected in relation to civil and criminal penalties. Table 7.1 shows only the most prominent and high-value settlements between major companies and the US government – there are others which are not listed because they fall below $300 million in headline value. In some cases, the company pleaded guilty to one or more charges. In others, a settlement was agreed without any admission of guilt on the part of the company. The scale of these penalties is staggering in a number of respects. First, because when the settlement of the first of these cases was announced in 2004, around Neurontin™, the general view was that the enormous size of the fine would be enough to deter subsequent companies from engaging in this practice. Clearly that was naïve: it has continued for a further 10 years. The Neurontin case, though a celebrity at the time, pales in comparison with the $3 billion settlement with GSK in 2012 (see Table 7.1). (But then again, as I have said all along, off-label practices are absolutely huge in scale.) The continuing litigation in this area therefore suggests that pharmaceutical companies have increasingly regarded this as just another cost of business associated with regulation, in other words, that the commercial gains from illegal off-label activities are worth the losses. Other commentators have remarked that US government sanction is of limited consequence besides the financial gains from illegal off-label marketing [231].

Second, the apparent nonchalance with which companies continued (and, perhaps, still continue) to engage with this practice casts these large organisations as poor citizens rather than purveyors of life-saving and enhancing medicines. Notably, one of the companies (Pfizer) appears three times in the table, for repeated offences: first in relation to Neurontin in 2004, then for a mixture of drugs in 2009 and finally

Table 7.1 Significant fines in relation to illegal off-label promotional activities of major pharmaceutical companies imposed by the US Department of Justice.

Company	Settlement ($m)	Violation(s)	Year	Product(s)
Pfizer	430	Off-label promotion	2004	Neurontin™ (gabapentin)
Serono	704	Off-label promotion/kickbacks/monopoly practices	2005	Serostim™
Schering-Plough	435	Off-label promotion/kickbacks/Medicare fraud	2006	Temodar™ (temozolomide) and others
Purdue Pharma	601	Off-label promotion	2007	Oxycontin™ (oxycodone)
Bristol-Myers Squibb	515	Off-label promotion/kickbacks/Medicare fraud	2007	Abilify™ (aripiprazole)/Serzone™ (nefazodone)
Cephalon	425	Off-label promotion	2008	Actiq™ (fentanyl)/Gabitril™ (tiagabine)/Provigil (modafinil)
Pfizer	2300	Off-label promotion/kickbacks	2009	Bextra™ (valdecoxib)/Geodon™ (ziprasidone)/Zyvox™ (linezolid)/Lyrica™ (pregabalin)
Eli Lilly	1415	Off-label promotion	2009	Zyprexa™ (olanzapine)
Allergan	600	Off-label promotion/kickbacks	2010	Botox™ (botulinum toxin)
AstraZeneca	520	Off-label promotion/kickbacks	2010	Seroquel™ (quetiapine)
Novartis	423	Off-label promotion/kickbacks	2010	Trileptal (oxcarbazepine)
Forest	313	Off-label promotion/kickbacks	2010	Levothroid™ (levothyroxine)/Celexa™ (citalopram)/Lexapro™ (escitalopram)
Merck	950	Off-label promotion	2011	Vioxx™ (rofecoxib)
GlaxoSmithKline	3000	Off-label promotion/kickbacks/failure to disclose safety data	2012	Avandia™ (rosiglitazone)/Wellbutrin™ (bupropion)/Paxil™ (paroxetine)
Abbott Laboratories	1500	Off-label promotion/kickbacks	2012	Depakote™ (sodium valproate)
Amgen	762	Off-label promotion/kickbacks	2012	Aranesp™, Epogen™ (erythropoietin)
Pfizer	491	Off-label promotion/kickbacks	2013	Rapamune™ (sirolimus)
Johnson & Johnson	2200	Off-label promotion/kickbacks	2013	Risperdal™ (risperidone)

Adapted from Wikipedia, List of largest pharmaceutical settlements (http://en.wikipedia.org/wiki/List_of_largest_pharmaceutical_settlements).

for Rapamune in 2013.[2] Third, most of the settlements involve fines for both off-label promotion and kickbacks or fraud. It could be argued that off-label promotion, though illegal, is merely the process of convincing someone else (i.e. a physician) to do something that is legal; however, the same thing cannot be said about the other two categories of offence, which I find gravely concerning to see mentioned repeatedly in Table 7.1.

Some of the details of the settlements are revealing. For instance, AstraZeneca's settlement in 2006 for illegal promotion of quetiapine was in regard to selling it for 'aggression, Alzheimer's disease, anger management, anxiety, attention deficit hyperactivity disorder, bipolar maintenance, dementia, depression, mood disorder, post-traumatic stress disorder, and sleeplessness'. According to my research, they may have left out further conditions which have been alleged to have been invented by pharmaceutical companies, such as binge eating disorder and body dysmorphic disorder. Many of these new conditions have been noted to have increased enormously in prevalence along with the availability of various off-label treatments. As far as quetiapine was concerned, at the time of the arraignment, the only licensed uses for this supposed cure-all were for schizophrenia and bipolar mania. Similarly, in prosecuting Abbott for alleged off-label promotion of the anti-epileptic Depakote (sodium valproate), the FDA quoted the following non-labelled uses: 'behavioural disturbances in dementia patients, psychiatric conditions in children and adolescents, schizophrenia, depression, anxiety, conduct disorders, obsessive-compulsive disorder, post-traumatic stress disorder, alcohol and drug withdrawal, attention deficit disorder and autism'. As I have said previously, these long lists of supposed applications are more reminiscent of the advertised (but unbelievable) applications of rhino horn, or ginseng, than 'ethical' pharmaceuticals.

Another case involved Pfizer's alleged promotion of the erectile dysfunction drug sildenafil (Viagra™) to treat low libido and to 'restore and increase orgasmic sensations' in women. Pfizer actually tried to develop Viagra for women, but their clinical trials failed; perhaps this is because women's sexual desire is more complicated than the mainly male executives at the company had thought when proposing this development. However, not all of the (alleged) breaches involved widely different indications. In some cases, the off-label indication was close to the approved label. In some of the antidepressant drugs in Table 7.1, the product was approved for adult use, but allegedly promoted to paediatricians specifically for young patients who demonstrated signs of depression. From this supposedly related use, it became apparent that suicide and suicidal tendencies were associated with use in young patients, whereas this is not the case for older patients (see p. 84). Finally, there are other cases in which the promotional activity involved different doses from those on the label. An example of this involved Novartis, the manufacturer of oxcarbazepine (Trileptal™) allegedly promoting use of the antiepileptic drug 'as monotherapy for seizures using extremely high dosages'.

During these cases, details also emerged about the methods used for off-label marketing. These included all of the activities previously outlined in Chapter 4. In addition, they included other methods such as direct financial incentives to physicians. Among these incentives were lavish gifts, honoraria or consultancy contracts with a promotional aim rather than to help the sponsor company obtain expert

---

[2]It is only fair to note, however, that Pfizer's appearance in off-label litigation both in 2004 and 2013 is as a result of corporate acquisitions it made. The earlier case arose from its takeover of Parke-Davis/Warner-Lambert, and the later case from its takeover of Wyeth.

advice. As well as these external activities, internal practices were adopted to conceal off-label marketing activities, to permit 'plausible deniability'. These included efforts to deliberately 'clean' internal communications and memoranda of references to off-label marketing. Success by individuals in off-label marketing was financially rewarded, since some of the sales targets were impossible to meet if the product was only sold with the regulatory label as a restriction. Finally, there were fraudulent activities related to reimbursement, with company representatives discussing with prescribers ways to bypass insurers' restrictions on prescriptions of the product, or how to falsify the billing code. Other methods involved gaining a mention for the off-label use on a hospital formulary or getting physicians to write letters in support of the off-label use. Companies were also engaged in bringing patients who might be candidates for off-label prescriptions to the attention of physicians. Sometimes the company would find indigent patients and offer them gifts in order to persuade them to seek out physicians for an off-label prescription [232].

In addition to governmental litigation applied for regulatory crimes and misdemeanours, pharmaceutical companies have been on the receiving end of civil suits from insurance companies too. Awards in these suits relate to claims that additional costs were incurred by insurance companies for reimbursement of off-label prescriptions. So, following on from the fine of $430 million imposed by the FDA in 2004 on Pfizer related to the company's improper marketing of the epilepsy drug Neurontin™ (gabapentin), the Kaiser insurance group embarked on a suit to reclaim some of the reimbursement costs associated with the off-label prescription use of the drug. This suit was litigated initially through a jury trial, then through various appeals right up to the US Supreme Court. Ultimately, Pfizer's appeal against a $142 million judgement was rejected [233]. As a result of the Supreme Court decision, similar claims brought against Pfizer with regard to off-label promotion of Neurontin™ by other insurers, as well as various state antitrust claims, were allowed to proceed. In June 2014, Pfizer reached a preliminary accord to settle out of court for a payment of a further $325 million. The accord followed another settlement from April 2014 in which Pfizer agreed to pay $190 million to resolve allegations that the company delayed generic competition for Neurontin™ [234]. When these latest settlements are added to the original FDA fine of $430 million, the total cost of fines to Pfizer for off-label promotion of Neurontin™ is at least $945 million. Presumably, now that the precedent has been set – and relates to one of the earliest off-label fines – we can expect similar claims will be brought by insurers against others in Table 7.1.

All in all, the catalogue of huge fines levied by the regulatory agencies has substantially damaged the public image of the pharmaceutical industry, with concerns raised about its preferences for shareholder value over the health of the patients they treat. In fact, one of the main stimulations for writing this book was to point out that this damage was wrought at the altar of off-label medicine, with the very large commercial attraction that the strategy brings. It is also true that the industry has relied on the medical profession as partners in this grey practice, yet physicians have not faced censure, even for the things which it is illegal for them to do. Optimistically, it could be that a more responsible period of industry practice is beginning. I note that since the most significant $3 billion fine was levied against GSK in 2012, GSK has emerged prominently in the vanguard of the reforming tendency, with an announcement in 2014 that it will no longer pay physicians to promote its products; it has also adopted other major reforms to its marketing policy and changes to its policy on data transparency. Behind the vanguard, there are also signs from other companies of genuine attempts to avoid the level of off-label marketing activity that pervaded the industry in the past two decades. However, I am not sure that GSK's lead will be eagerly followed by all other companies. Some companies may prefer to adapt to the regulatory environment and maximise their commercial profits within it, rather than worry unduly about the public relations damage of civil suits. If faced with more stringent regimes in future, including possibly custodial sentences, there is a danger that rather than change corporate course from what is a very profitable activity, they will adopt procedures which avoid paper trails and promote 'plausible deniability'. It also turns out that there is a legal limit to the controls imposed by regulators on marketing activities, as discussed later (see p. 139).

Whistle-blowers

A main reason the US authorities have been so successful in prosecuting this behaviour is because they have been able to access first-hand accounts from industry insiders, known as whistle-blowers. Whistle-blowers are private individuals (known legally as *qui tam* witnesses) empowered to bring actions on behalf of the government, and they may receive up to 30% of any recovery. More than 900 whistle-blower suits were filed in 2012, of which around 10% involved pharmaceutical companies, and a small number of those ended up attracting the attention and backing of the Justice Department. In 2011, *qui tam* witnesses earned more than $558 million in share awards, which is the highest yearly recovery on record and relates to settlements of around $2.8 billion. For example, it was reported in 2010 that five whistle-blowers would be awarded a total of $38.7 million out of the $600 million Allergan settlement for off-label promotion of Botox™ [235].

Another tool the FDA has used to encourage adherence to its rules is the so-called 'Bad Ad' programme. This educational outreach programme is designed to educate healthcare providers about the role they can play in helping the agency make sure that prescription drug advertising and promotion is truthful and not misleading. Examples of violations include omission of risk, promotion of an unapproved use or overstatement of effectiveness. The activity can be through a presentation at a seminar, an educational event or any of many other kinds of promotional event that companies organise. The FDA also asks physicians to report these violations.

European situation

By comparison to the activities in the United States, efforts to control this activity in Europe have been puny. This is partly because the responsibilities are subsidiary to the EU itself and are devolved to the individual states, which have different rules. The recent introduction of Regulation 658/2007 in the European Union changes this situation and empowers the EU Commission to impose financial penalties for corporate violation of EU legislation on medicines, but this is rather late in the game since the alleged crimes have been going on for many years [236]. Differences between Europe and the United States also arise because in many European countries, pharmaceutical companies need to negotiate with government regarding pharmaceutical prices; as we shall see, this means government may have another non-litigious means to control malpractice but one with substantial leverage. One of the more committed European countries in regard to legislation to control off-label marketing is France, where the Mediator scandal (see p. 102) caused substantial political ructions (as well as patient harm) and resulted in an accord between the government and the pharmaceutical companies in December 2011 [237]. In this accord, there was a commitment to limit the extent of off-label use, to ban the advertising thereof and to put in place measures to make sure medicinal use complied with the recommendations and an announcement of a financial penalty in case the companies failed to comply with the agreement. So far, such a penalty has failed to arise, and there are beginnings of a new regime to regulate off-label use, which will be described in detail in the last chapter.

In the United Kingdom, the regulation of marketing practices falls not to the pharmaceuticals regulator, the MHRA, but to the Prescription Medicines Code of Practice Authority (PMCPA), a quasi-autonomous branch of the industry lobby group, the Association of the British Pharmaceutical Industry (ABPI). The ABPI has the ultimate sanction for companies that breach the ABPI code (which includes advertising) in the ability to rusticate offenders and thereby expel them from an ability to negotiate with the UK government on matters related to pharmaceutical pricing. The process works through self-regulation, and cases alleging contravention can be brought in an open forum by anyone with an interest. The open nature of the confrontation enables clarity about what is and what is not allowed. This is important because the US experience has been rather plagued by accusations and counter-accusations that the FDA rules are unclear, and also because the dividing line between what is promotional and what is factual is (even in the MHRA's words) 'not always clear'. The process has not been without its critics, and in 2004–2005, a UK Parliamentary Select Committee on Health censured the working of this process as well as the practices of the industry. As a result of this criticism, substantial improvements were made; these were driven both by the industry who were anxious to retain the advantage of the system of self-regulation and by the MHRA who were under political pressure to respond to the report and reform the system.

The PMCPA process enables anyone with a grievance, drawn from competitors, former employees, physicians, patients and the MHRA itself, to bring complaints against ABPI members for violating the advertising and promotional rules and regulations. Complaints may be initiated anonymously, to protect the complainant. The case is rapidly adjudicated, and the reasoned outcome posted on the website. The medical and pharmaceutical knowledge of the panel adjudicators represents a fairly high level of sophistication and judgement. However, it is also apparent that decisions tend to give the benefit of the doubt to the industry defendant. For instance, an allegation was made in 2012 against GSK for improperly promoting the use of Revolade™ (eltrombopag) for myeloid fibrosis, a disease of the bone marrow for which the drug had no licence; the MHRA found that an individual company representative had breached the code. However, it did not extend culpability to GSK itself, finding the company not guilty of the more serious charge of bringing discredit upon the pharmaceutical industry. This individual case suggests a less rigorous enforcement regime in the United Kingdom relative to that operated by the US authorities, but a recent opinion from John Osborn, associate professor at the University of Washington School of Law, is complimentary of the UK system relative to its more adversarial US counterpart [238].

In addition to the issue of pricing negotiation, the UK experience also has another important difference relative to the US one. That is, for most prescriptions that a UK doctor writes, there is no limitation imposed on reimbursement according to the intended use. A UK prescription to a demented 92-year-old lady to control her aggression in a nursing home is reimbursed in the same way as a prescription to a 31-year-old schizophrenic in a secure mental hospital. In the United States, the reimbursement of most off-label uses is a crime under the False Claims Act, and this has been part of most of the settlements in Table 7.1.

There are no comparative data on one of the most disturbing components of the settlements in this table: that of kickbacks and Medicare fraud. As mentioned earlier, this concerns corrupt payments to physicians by the pharmaceutical industry for off-label prescriptions and false reimbursement claims therefrom. It is possible though uncertain that the greater laxity of the UK reimbursement regime, relative to the United States, may have had a role to play in this too. What I mean by this is that the way that drugs are reimbursed in the United States often requires such prescriptions to be associated with a false reimbursement claim – for instance, in situations where Medicare does not automatically reimburse for the use of a particular drug in a particular condition. Insofar as US off-label reimbursement involves a crime, a US physician may feel he needs to be compensated with kickbacks in a way that was not relevant for the UK doctor, where, as described in the preceding paragraph, the NHS reimburses in all cases.

Nevertheless, the law in the United Kingdom has now changed so that under the 2011 Bribery Act, companies headquartered in the United Kingdom can now be prosecuted for acts of bribery that occur anywhere in the world. Claims may be brought not just for acts of bribery *per se*, but for a failure to prevent such acts. There have been recent events in China, Iraq and Poland involving allegations of corruption and bribery which may be used to bring cases against GSK in this regard. A further tightening of the acceptable marketing practices across the UK industry also took place through the pharmaceutical industry body, the ABPI. In 2011, this group also changed its policy so that companies may no longer offer promotional aids such as stationery items, computer accessories and so on to doctors (previously they had been allowed if they were low in value, now they are not allowed even if cheap). The ABPI code is important because if abided by, it may be used by a company in its defence against a claim under the Bribery Act [239].

But marketing practices aside, we need now to look at things from the patient's viewpoint and ask how to build a regime for optimal patient benefit, not just for apportioning blame.

Tip of the iceberg

It is fairly clear that the main purpose of regulatory involvement in off-label promotion is the protection of the public through inappropriate off-label prescriptions. But if this is the case, even if the manufacturers of branded pharmaceuticals represent the low-hanging fruit, there are plenty of other instances where inappropriate off-label prescriptions occur that are not affected by this form of regulation.

The restraints on branded off-label promotion do not extend to generic producers, mainly because these companies do not promote their products. And, as we dealt with earlier, generics represent around 85% of prescriptions. In terms of looking at the problem as an iceberg, most of the dangers to patient health lie hidden beneath our superficial gaze.

The regulatory solution based on attacking off-label promotion can only ever be a partial one: the regulators may do their best to prevent these activities, but promotional efforts only exist for drugs still on patent. After the patent has expired, the damage created by off-label promotion has not gone away; doctors still have in their minds that drug X can be used for indication Y or in a different way from that authorised by the regulatory authorities. Their attitudes have been imprinted by a set of medical myths which pervade prescribing practice; the scientific literature remains corrupted by the same publication bias, flimsy evidence, ghost-written articles and so on that filled promotional leaflets of a bygone era. Off-label promotion may have gone, but the unexploded mines still litter the medical practice landscape.

Generic companies do not promote; they devote all their activities to copying others' products. So they are immune from the current FDA approach to off-label regulation. But they still sell products which are deficient in evidence in the same way as their innovator predecessors did. Moreover, since the drugs are cheap, there is even more incentive for the prescriber to prescribe them. At this point in the product life cycle, the opportunity to invest in investigating the further therapeutic utility has gone: there is no longer any commercial incentive to prove the off-label use works (even less to prove it doesn't). We are in an era where any safety issues associated with the off-label use are likely to be diluted by the on-label use, unless we can find ways of categorising and delineating safety and therapeutic use at the same time. In the generic period, unproven off-label uses persist and become embedded in medical practice, accepted unchallenged until a publicly funded exercise unearths the dangers.

How much evidence is there that this really goes on? Well, here are three snippets. We saw that Cephalon's sales of Provigil™ continued to increase after their legal settlement in 2008, with a huge increase in sales after the settlement had been reached in 2008–2009 [55]; in other words that the cessation of off-label activities did not immediately bring down the level of off-label use – in fact, the reverse. A similar situation was argued to have occurred with Neurontin™, off-label use continuing after patent expiration; this was litigated with respect to a patient prescribed with generic gabapentin for leg pain, and who developed suicidal symptoms and severe depression thereafter [240]. Second, we also know that drugs continue to be used off-label without good evidence for a long time and are only displaced once someone bothers to evaluate their supposed benefits; this can take many years. For more than 40 years, corticosteroids were routinely used for the treatment of brain trauma, under the non-evidenced assumption that they alleviated brain swelling; yet these drugs were not promoted for this use, and when evidence was finally gathered, it was shown that they were actually damaging in such patients, causing an extra 2.5 deaths per 100 treated [241]. In fact, this is not the only area where steroid use has been criticised: the long-standing tradition of using these drugs in paediatric septic shock has also been questioned as an area of evidence-deficient off-label medicine [242]. The third point is that if off-label promotional activities, which have by common assent risen in prominence in the past two decades, were a primary cause of increases in off-label prescriptions, there would be more off-label use among new drugs relative to old drugs. In fact, the opposite is the case. In the study by Eguale referred to earlier [43], drugs approved after 1995 were prescribed off-label less often than were drugs approved before 1981 [43]. When combined, the evidence makes it look as though the regulatory police are chasing the wrong man. Generic medicine is highly implicated in off-label patient harm, yet all the regulatory attention is focussed on branded products.

This is not to say that generic medicines do not benefit from increased evidence becoming available on branded drugs. The story of paroxetine and suicidal behaviour in young people taught us something about many of the traditional ('tricyclic') antidepressants, which now also carry a regulatory warning about their use in younger patients being associated with suicidal behaviour. This information derives from the detailed work done on a branded drug (paroxetine) in a particular patient group. For many years prior to this revelation, tricyclic antidepressants were given to young and old without concern for a difference in outcome. Unfortunately, it is not in the commercial interests of generic companies to undertake such detailed work on their products.

Meanwhile, in comparison to the persistence with which established, non-evidenced off-label use may remain obstinately embedded in medical practice, the adoption of off-label uses even in a new agent can be very rapid indeed. Dabigatran (Pradaxa™) is a relatively new anti-clotting drug approved by the FDA in 2010. Within 12 weeks, it had been associated with over 300 serious adverse events, with a high preponderance of off-label use. Although only approved for the prevention of stroke in patients with atrial fibrillation, it had become widely used for the treatment of a range of general but unspecified purposes related to inhibiting blood clotting, amounting in total to over 60% of the uses for the drug. The data show the speed with which a new treatment can spread into wide clinical use, generating reports of hundreds of serious injuries within weeks. It is troubling that this new anti-clotting drug was immediately used off-label where its risks and benefits had not yet been systematically studied [243].

We saw earlier that a large proportion of medical myths are retained without formal proof of their inutility (see p. 101), suggesting that it is only by a prolonged and active campaign of evidence gathering that we can rid ourselves of exposure to these practices. One way in which this is starting to occur is through pharmacovigilance: the process of assessing pharmaceutical safety through regulatory supervision of adverse events and a process which is undergoing radical overhaul and expansion right now. We shall discuss this further in Chapter 8; before doing so, it is important that we now end our discussion of regulatory control of off-label promotion with a fascinating analysis of how this interacts with the conflicting precepts of free speech.

Free speech

Given the importance of the constitutional right to free speech embodied in the First Amendment to the US constitution, regulators have been wary of interfering too deeply into commercial advertisements. We saw earlier how the United States is almost the only country to allow direct-to-consumer pharmaceutical adverts and how this arose from a consideration of the right for commercial organisations to advertise; pharmaceuticals (unlike tobacco companies) were no exception.

Despite this freedom, there was also concern that unfettered free speech for pharmaceuticals might damage public health because patients might be misled and thereby exposed to harmful or inefficacious products. Therefore, an intermediate standard prevailed; it was called the Central Hudson test. It derived from a case in which the New York Public Service Commission had tried to ban electricity companies from airing adverts promoting the use of electricity. The utility company Central Hudson challenged the constitutionality of the law and ultimately prevailed in a Supreme Court judgement of 1980. The precedent established the Central Hudson test, which was interpreted as far as pharmaceuticals were concerned to allow governmental interference in commercial free speech provided the regulatory restrictions were at the minimal level necessary to directly advance public health.

The Central Hudson test underpinned a change of the FDA rules during the 1990s, a time during which the agency also increased restrictions on tobacco advertising. Emboldened by their success in this area, the FDA continued to mount actions against off-label promotion, during which it became clear that the huge financial consequences of each settlement hinged significantly on the allegation that the wrongful acts committed by the companies involved 'false and misleading speech'. After all, the very premise of the regulatory restrictions was to prevent patients being 'misled' into unsafe or ineffective treatments.

But what if the marketing claims were not false; what if the speech was 'true'? Where off-label prescribing lacks efficacy or poses safety problems, there is a legitimate, compelling interest in protecting public health by ensuring that companies do not mislead or otherwise encourage it. But where the challenged off-label information is truthful, surely the converse is true, and its promotion ensures that physicians are better informed and able to provide good medicine. What is the public interest in preventing this activity? This was not just a moral question, but a matter of constitutional right. The First Amendment protects not only the right of the individual to free speech, but also the right of commercial organisations. This protection applies so long as it is 'true'. Is it therefore right that the US government, in the guise of the FDA, seeking to prevent off-label marketing, prosecutes pharmaceutical companies regardless of whether that marketing is true or false?

During the 1990s, it became established through a series of law suits that the FDA did not have the right to prevent the dissemination of truthful, non-misleading scientific and medical information, at least in the form of peer-reviewed journal articles, medical textbooks and sponsorship of continuing medical education programmes. The possible use of this line of defence by pharmaceutical companies grew as the off-label lawsuits grew in number and in size. For instance, Neurontin™, which was the subject of the first celebrated off-label suit against Pfizer, was indeed eventually properly developed for neuropathic pain, showing that it was safe and efficacious for the main indication at the heart of the $430 million fine. By the time that the regulatory approval had been granted, the fine had been paid and there was no going back, but a philosophical question remained: how could its promotion in this way qualify as a 'false and misleading', when the actual medical utility proved otherwise? Then in 2010, a prominent suit involved Allergan's promotion of onabotulinumtoxin A (Botox™) for unapproved indications including headache, pain, muscle spasticity and cerebral palsy in children. One of Allergan's defences was a First Amendment challenge against the regulation prohibiting off-label promotion. Ultimately, Allergan pleaded guilty to a misdemeanour misbranding charge and paid the US government $600 million, but in announcing this settlement, the company expressed regret that the First Amendment issue had not been litigated to a final conclusion. As we know from Chapter 2, Botox is now approved for 10 labelled indications including conditions like chronic migraine and upper limb spasticity.

Although FDA regulations warn that it is considered 'misbranding' for marketeers to 'recommend or suggest' that a drug is appropriate for an indication for which it has not specifically been approved, the act the marketeers seek to encourage is legal for physicians to do. In a 2011 decision around the use of pharmacy data for pharmaceutical marketing purposes (*Sorrell v. IMS Health*), a majority of the court concluded, 'If pharmaceutical marketing affects treatment decisions, it does so because doctors find it persuasive', and 'the fear that speech might persuade provides no lawful basis for quieting it'.

The case law established by *Sorrell v. IMS Health* and the settled suit involving Allergan finally set the scene for an even more important case involving Alfred Caronia, a sales representative of Orphan Medical, a subsidiary of Jazz Pharmaceuticals. The case concerned the promotion of Xyrem™ (sodium oxybate), a central nervous system depressant approved for the orphan-designated treatment of excessive daytime sleepiness and cataplexy in adults with narcolepsy. Cataplexy (as you may remember from earlier) is the sudden loss of muscle control associated with an involuntary urge to sleep. The active ingredient in Xyrem is the sodium salt of gamma-hydroxybutyrate, also known as 'GHB' or the 'date-rape drug' for its role in many sexual assault cases, as well as being commonly used for consensual recreational purposes. So, similar to opiates, benzodiazepines, amphetamines and ketamine, which have legitimate and illegitimate uses, the chemical has twin personalities, being approved as a pharmaceutical for prescription medical purposes, as well as being classified as an illegal drug in the wrong hands.

Even though the FDA had just approved Xyrem™ to treat rare conditions, Orphan Medical was interested in wider application. The commercial opportunity from off-label use was huge, and we have seen previously how companies use the orphan drug legislation as a platform for much wider, non-orphan use of their products. In the United States, the cost of Xyrem equates to between around

$4600 and $9250 per month (prices as of October 2013); costs in the United Kingdom are much lower, around 10–12% of this level. When the federal government launched an investigation into the promotion of the drug, it trapped one of the company's sales reps, Alfred Caronia, in a sting operation for promoting Xyrem for a variety of other uses. Here's an excerpt from one of Caronia's conversations with Dr. Stephen Charno, who posed as a potential customer:

> [CARONIA]: And right now the indication is for narcolepsy with cataplexy...excessive daytime...and fragmented sleep, but because of the properties that...it has it's going to insomnia, fibromyalgia[,] periodic leg movement, restless leg, ahh also looking at ahh Parkinson's and...other sleep disorders are underway such as MS.
> [CHARNO]: Okay, so then ... it could be used for muscle disorders and chronic pain and...
> [CARONIA]: Right.
> [CHARNO]: ...and daytime fatigue and excessive sleepiness and stuff like that?
> [CARONIA]: Absolutely. Absolutely. Ahh with the fibromyalgia.

Caronia worked with a psychiatrist in private practice (Dr. Peter Gleason) on a consultancy basis to help proselytise the wider uses of Xyrem, so that Gleason gave speeches and visited other physicians to discuss off-label uses of the drug. They were caught promoting the drug for use in children under age 16, which again was outside the labelled use. For the record, Xyrem is indeed prescribed off-label for fibromyalgia, depression, schizophrenia, chronic fatigue syndrome and severe cluster headaches.

Caronia was charged with criminal behaviour under the 'misbranding' provisions of the FDCA and was convicted at trial. Elsewhere in the litigative process, Gleason pleaded guilty to a misdemeanour charge, and Orphan Medical agreed to a $20 million criminal and civil settlement. Gleason's medical licences were suspended, and his life fell apart. He committed suicide in February 2011.

Gleason's tragedy was perhaps the greatest of the three strands of litigation. Orphan Medical had no overt commercial decline from the fine it had to pay; but the most interesting outcome came from the criminal suit against Caronia, since he did not meekly accept the first verdict. He appealed that conviction on the grounds that criminalising off-label promotion violated his First Amendment right to free speech, and he prevailed with this argument in a Second Circuit judgement. Drawing on the Supreme Court's determination in 2011 in *Sorrell v. IMS Health* that '[s]peech in the aid of pharmaceutical marketing...is a form of expression protected by the Free Speech Clause of the First Amendment', two members of the three-judge panel concluded that drug companies and their representatives cannot be prosecuted 'for speech promoting the lawful, off-label use of an FDA-approved drug'.

In reaching that conclusion, the court found that the FDA rules on what a sales rep could say to be unclear, ambiguous and too broad and wrong that drug companies could be prosecuted for promoting off-label uses when those uses are themselves lawful. At one point, the prosecution tried to explain that off-label promotion 'is not a crime' *per se*, but is merely evidence of Caronia's and Gleason's intent to 'introduce a misbranded drug into commerce', which is illegal. But the only way the drug was 'misbranded' was Gleason's claim that it was safe and effective for off-label uses. The judges had difficulty following the prosecution's argument and questioned why such speech should be considered illegal.

Ultimately, the court took the position that it would be inequitable for the company to be prevented from saying the same thing that a doctor, in the reverse position, could legally say. One of the three-judge panel dissented, finding that the free speech argument should not allow companies to run around the law of drug regulation that underpins the FDA. In her view, the prosecution was for a conspiracy to sell a drug without approval, and the speech is simply part of the evidence of that conspiracy, without which the conspiracy cannot be proved.

The FDA decided not to appeal the decision of the Second Circuit to the Supreme Court, presumably concerned about the impact of a negative outcome. It therefore remains as binding law on courts within the Second Circuit and as a potentially persuasive precedent in other courts. Even though we are left with a judgement that has not been tested at the highest level, it poses serious challenges to the FDA policy of prosecuting off-label marketing and may even call into doubt previous settlements made by pharmaceutical companies since the start of the new millennium. In May 2014, the FDA's Chief legal counsel said the organization was "…carefully evaluating … [its] … policies in light of court decisions on first amendment issues" [244]. In the ensuing short period since the judgement was handed down, there are already signs that the US government is reluctant to take on cases against off-label pharmaceutical promotion unless they involve additional factors such as corrupt payments, safety issues, manipulation of clinical study data and so on.

Critics of pharmaceutical marketing will however be especially alarmed at this decision; not only might it give much greater freedom to off-label marketing campaigns, and effectively curtail the ability of the FDA to police them, but it is particularly troubling that the product in suit is based on a street drug with illegal overtones. Arguably, if the law of free speech protects the off-label use of a drug with these connotations (even though not specifically in these ways), what chance is there of regulating the egregious promotion of the next antipsychotic or anticoagulant medicine?

In order to understand this decision properly, it is important to note that it is not all about free speech. While US law protects the right of the pharmaceutical marketeer to promote outside FDA regulations, the success of Orphan Medical is derived as much from the laxity of off-label prescribing. The company prevailed precisely because it argued it was not asking prescribers to do anything illegal. An off-label prescription can be written, from a legal perspective, almost identically to an on-label prescription. And the AMA says almost nothing in its professional guidance to challenge that freedom.

A trite solution to this kind of marketing practice would be to curtail the physician's right to freely prescribe off-label, since this would also curtail the free speech defence of these marketing practices. Thus, should off-label prescription be illegal, marketeers would not be able to promote towards this end. We have seen earlier in this book why the blanket prohibition of off-label medicine is not a good idea, but nevertheless, one of the consequences of the present situation is the widespread off-label use of products like Xyrem.

The second point is about the 'truth' of the promotional statements. From the FDA regulatory position, the use of Xyrem was approved only for excessive daytime sleepiness and cataplexy with narcolepsy. But from the physician's perspective, a prescription can be written with much lower levels of proof. We have seen in Chapter 6 that physicians need only to be able to justify their actions in accordance with a respectable, responsible body of professional opinion. One or two peer-reviewed articles about Xyrem in these non-approved areas should suffice to support the 'truth' that the product can be used in these conditions. But what quality do these articles need to adhere to? Who is to judge whether they are subject to publication bias? Might they have been ghost-written by a pharmaceutical marketing department and merely labelled with key opinion leaders who lend their name and credibility to the article without significantly contributing to the content? Are the trials robust and well conducted, or small unblinded studies with protocols designed to favour positive outcomes, and conclusions that over-egg the results? What chance would this evidence have of being accepted by a regulatory authority or other independent expert panel? It is this gap, between what is required to convince the regulators, and what is required to convince a physician, that is being played out by Orphan with Xyrem.

This is a terrible mess. While campaigners cavil at the devious means pharmaceutical companies use to avert the prosecutional attention of the FDA, that same FDA is prevented from more stringent action both by the immovable standards of free speech and the immutable determination of the medical profession to resist off-label regulation. However, all is not lost; there are ways, as we shall discover in Chapter 8, of reining in some of this practice, but we do need to consider new approaches, different from those of the past, to put a brake on the over-prevalent practice of off-label medicine.

CHAPTER 8

Justifying unapproved medicine

Let us summarise where we are: off-label medicine has its uses, but the principle is not universal. Thalidomide was recalled from the scrap heap and reinvented as a treatment for leprosy and multiple myeloma. Originally, it was used to treat insomnia. The principle of off-label use, if universally good and applied universally, would allow the reintroduced thalidomide to be used again for the treatment of insomnia: this after all was the original use for the drug, before it was withdrawn, so we know it will work! This is clearly ridiculous, since such an allowance would expose patients to unwarranted dangers given the non-seriousness of the condition, insomnia. The adverse effects of thalidomide are not limited to damage to the unborn child: the drug also causes a wide variety of other side effects including peripheral neuropathy, or tingling or painful sensations in the hands and feet, and sensory loss in the lower limbs. It can also increase the risk of serious blood clots in the legs or lungs, as well as heart attacks and strokes. All these effects occur both inside and outside pregnancy.

We can also look at the inconsistency between regulation of medical practice and regulation of promotional activity. While off-label medicine is legal, and suggested to have some benefits, its promotion is illegal because it is suggested it would lead to undesirable levels of off-label prescriptions. Clearly, if off-label medicine is a good thing, it is a good thing we do not want too much of. This is not a sound starting point for health policy. How can we tell when an off-label prescription is good medicine and when it is bad?

Or we can look at it from the point of view of regulation of pharmaceuticals. When a new drug is approved, the regulators carefully assess both safety and efficacy in respect of the proposed therapeutic use. But once approved, every subsequent use is allowed on the basis of presumed safety and without regulatory assessment of efficacy. What is it about a drug's second use which allows us to accept a different regulatory standard from that applied in its first use?

Whichever angle it is viewed from, we have a problem.

Off-label prescriptions run on average at the rate of one in five of all prescriptions, of which there are billions worldwide, so it is not a trivial issue. Adverse events from off-label medicine run at around two- to threefold that of drugs prescribed on-label. Noting the number of overall deaths associated with drug-related adverse events from page 87 (106 000) and assuming a relative risk factor of between two and three-fold, we calculated in Chapter 5 that there are around 35 000–45 000 deaths per year associated with off-label drug use in the United States. On a pro rata basis, using the figures quoted on page 88, this equates to a figure of approximately 100 000 deaths annually across the OECD countries. This is a substantial amount of patient harm for a group of medicines of which three quarters have little or no good scientific evidence for their benefit.

We need to do more to justify this situation. A lot more.

Off-label Prescribing – Justifying Unapproved Medicine, First Edition. David Cavalla.
© 2015 John Wiley & Sons, Ltd. Published 2015 by John Wiley & Sons, Ltd.

Three trends in medicine make off-label prescriptions an issue where yesterday's practice is not acceptable. First is the increasing demand for safety in the drugs we take. The balance of benefit and risk is slanting evermore towards a concern for risk over and above a yearning for benefit. Second, the patient–doctor relationship has changed enormously since the end of the Second World War. Doctors are still held in respect, certainly more so than the pharmaceutical companies who make the medicines, but the respect is qualified. The reason is partly because patients are better educated and have better access to information but also because their expectations have changed. Informed consent is critical.

And the third trend is the increase in evidence as a basis for current medical practice. The old shibboleths are being challenged as experimental trials disprove the prevailing standards of care. Drugs are approved by regulators with huge accompanying data packages, leaving previous regimes of medical treatment looking alarmingly naked in comparison.

Faced with these challenges, there are a few necessary changes.

Constraints on making changes

Before going further, let us briefly consider the constraints against changing the current system.

Unfortunately, the freedom to prescribe freely in an off-label fashion is considered by many physicians to be a dearly held right, and changes thereto have met with profound resistance from the medical professions. In 1972, for example, the FDA announced its intent to reign in on off-label practice, saying it was obligated to act 'when an unapproved use of a new drug may endanger patients'. Thus, it proposed to consider revoking the approval of any drug extensively used off-label, regulating off-label uses as experimental (just as if the drug was a new drug) and limiting distribution channels to hospitals or physicians with special qualifications. The medical profession, including the American Medical Association (AMA), objected vociferously, and the FDA backed down. The medical profession sees government restrictions on medical practice as totally unacceptable and has successfully lobbied the US Congress to refrain from governmental interference ever since.

Nevertheless, in other European countries, medical professional bodies have worked cooperatively with regulators. We have professional guidelines in the United Kingdom and Germany, for instance, which discourage off-label prescriptions if there is an on-label alternative and reserve off-label medicine for the most serious diseases. These guidelines seem sensible compromises that reflect the increased risks associated with off-label medicine but recognise that none of the well-worked examples in Chapter 2 could have happened without an innovative component and the necessity of this course on some occasions.

An even more vigorous opposition would be fielded against moves to ban the practice of off-label medicine; in addition to the medical professions, one may also expect opposition from the pharmaceutical industry which derives significant incomes from this practice, both for branded and for generic products. In a survey completed by around 500 physicians, researchers at George Mason University in the United States found an overwhelming response against the banning of off-label practice, with 94% in opposition to such a move [245]. The respondents were mainly in favour of the current regulatory requirement that approved medicines require proof of efficacy for their first use, but (somewhat inconsistently) against a requirement that approved medicines require proof of efficacy for their second and subsequent uses.

In addition to this inconsistency, there was a substantial minority of doctors who agreed with a relaxation of the current rules to allow medicines to be used without proof of efficacy (which was the situation in the United States prior to thalidomide).

The physicians who argued for retaining proof of efficacy at regulatory level but not for proof of efficacy for an off-label prescription based their argument mainly on the idea of the off-label prescription being 'closely related' to the labelled status. Thus, 'children are like little adults', or 'all antihistamines are alike', or 'atrial arrhythmia is similar to ventricular arrhythmia'. We have dealt with these questions in Chapter 5. If all drugs within a certain class were indeed similar, that would suggest proof of efficacy for the second and subsequent medication that fitted within a certain class could be skipped in a regulatory review. If operative, Pfizer would not have had to prove that Lipitor™ reduced cholesterol levels because Merck had already had approval for their statin drug. I doubt such a proposal would meet with much acclaim outside the medical community. We have also dealt with the example of amiodarone, which is indeed useful in both atrial and ventricular arrhythmia, but associated with life-threatening side effects that do not justify its use in the former indication. Moreover, not all off-label medicine is closely related to on-label situations. How close are pain relief, prevention of stroke and prevention of cancer; yet all, with varying levels of proof, are treatable with aspirin? The unrelatedness of much off-label medicine is acknowledged by both those in favour and opposed to it.

Moves to enhance off-label medicine

The essential dilemma of regulation is shown in Table 8.1; a perfect system is able to avoid both the approval of bad drugs that were thought to be good and the denial of approval of good drugs that were thought to be bad. These are the true positive and true negative outcomes from a perfect determination that can never be. The size of a clinical trial to identify a safety hazard depends on both the frequency with which the drug causes the adverse event and the background rate of that event. For instance, we would need 160 000 patients in a trial to detect an excess drug risk of one in a thousand for myocardial infarction among middle-aged men [246]. This is completely impractical. In reality, a system that waits until all the risks of a new drug have been identified causes patients to be harmed by the disease while the full assessments can be carried out; it also makes the economics of pharmaceutical R&D relative to our ability to pay for new products completely unworkable. On the other hand, a system that responds rapidly and approves new products quickly risks harming patients because the drugs are unsafe. When considered in this way, regulatory decisions are no longer a balancing act between benefits and risks, but a trade-off between different kinds of risk – in other words, the risk from untreated disease relative to the risk from the drug itself [247].

Once approved, as we know, the drug is often used in unintended ways that are not controlled by the regulatory agencies, who see off-label use as problematic. This in itself engenders caution in the approval process, with consequent delay in the treatment of patients by a product that, absent concern about its off-label use, would be approved

Table 8.1 The dilemma of regulatory drug approval.

	Actuality: good drug	Actuality: bad drug
Regulatory opinion: good drug	Approve	Patient harm from drug
Regulatory opinion: bad drug	Patient harm from disease	Deny

without hold-up. The problems with a long regulatory review process were vividly brought to light with the AIDS epidemic of the 1980s, and under strong pressure from patient advocacy groups, regulatory approvals were advanced in order to offer therapeutic options more rapidly for a disease which equated therapeutic delay with death.

One of the limitations on measures to quell extensive off-label medicine is the desire by some patients for a regulatory system that permits flexible medical solutions to their disease. In other words, some patient groups would prefer to see greater freedoms for doctors to prescribe products with only limited evidence, particularly in cases where the treatments are for serious illness. As mentioned earlier, Lord Saatchi attempted in 2013 to introduce a private member's bill (the Medical Innovation Bill) to the House of Lords with the aim of improving access to life-saving alternatives in cases like cancer (his wife died of the ovarian kind). Though ultimately unsuccessful in garnering government support, by April 2014, the bill was in the midst of an unfinished consultation process, and its significant support from various quarters is evidence of the public desire for drug regulation to be permissive for serious illness.

Another theoretical approach to limit the extent of off-label medication is to expand the label that is granted by the regulators in the first place. With a wider label, physicians would not need to prescribe real-world patients with drugs that are not specifically licensed for the diseases they suffer from. We are however limited here by two factors: regulators are cautious, and the data supplied to them often quite specific in the types of patients and disease being treated. When these two factors are combined, the result is a regulatory label that is quite narrow. We have seen the effect of this, for instance, in our discussion of Neurontin™. The approved label for the product during Warner-Lambert/Pfizer's off-label marketing campaign was for adjunctive treatment of epilepsy: in other words, 'add-on' treatment. The product was not approved as monotherapy, and of course not approved outside epilepsy, even though so sold. The reason it was not approved as monotherapy is because it was not used as monotherapy in the trials that were conducted for regulatory approval: in practice, when you have a *potential* new treatment for epilepsy, you want to avoid situations where the patient is withdrawn from their existing treatment. Patients are recruited who are inadequately treated with one agent, and the new treatment added in to the regime. The trial results then compare the effectiveness of adding the new treatment relative to adding a placebo to the existing anti-epileptic therapy. No information is acquired (at least initially) for the effectiveness of the new agent alone, so it is only labelled as 'adjunctive'.

Proposals for expanding the initial regulatory label are therefore unlikely to succeed. The limitations imposed by a specific set of approved uses are generally there for a good reason.

Diagnosis shifting

Off-label medicine that derives from an unapproved use is categorised as such based on the diagnosis the doctor places on the patient. At some stage, honesty in prescribing requires the doctor to confirm the off-label nature of the medicine. However, some practitioners may react against an onerous set of restrictions on off-label medicine by 'diagnosis shifting' as described on page 11. When behaving in this way, a doctor would write down a diagnosis that complied with the approved label, even though the patient's actual diagnosis would correspond with a different scenario. Similarly, pressure brought through reimbursement practices, or concern about legal liability, can bring about this kind of behaviour. In some cases, there is evidence that pharmaceutical companies have assisted physicians to fill in the reimbursement forms with an incorrect diagnosis code, which was alleged in the Xyrem™ case [248].

This type of practice is, of course, unethical and may in some circumstances amount to fraud on the reimbursement authorities. It may give an insurance company a reason for invalidating a physician's medical insurance policy and reveal further litigative exposure. Professional sanction may be brought to bear as well, as it suggests the physician is not being honest with the patient about their condition. Nevertheless, at the margins it may occur, and the more legal restrictions we place on medical practice, the more likely this is to become a problem. An example of how this can occur, for different reasons, is the prescription of oral contraceptives to teenage girls in order to 'regulate their menstrual cycles', rather than to face their parents' wrath with the revelation that they are having underage unprotected sex.

A second issue around diagnosis is that of 'therapeutic creep'. This is the situation where a product that is on the market for a particular condition becomes used for a primarily related condition and then for a secondarily related condition (i.e. related by two 'hops' from the approved indication). The way in which factor VIIa (NovoSeven™), approved only for a rare case of haemophilia, became used to promote clotting in a wide variety of conditions is an example of this; so is erythropoietin, which was approved for the treatment of anaemia in patients on dialysis but became much more widely used for anaemia generally. In both cases, patients suffered excess strokes and heart attacks from these practices and were denied effective therapy. Physicians have a duty to be vigilant regarding this phenomenon and to avoid being enablers of the scientifically unsound transformation of prescribing conditions [249].

In order to segregate the 'good' off-label medicine from the quackery, we need more evidence.

A partial solution: clinical trial transparency

In the United Kingdom in 2013, the AllTrials movement was established to shine a spotlight on the phenomenon of hidden clinical trial data, whereby pharmaceutical companies and academic investigators failed to publicise fully the results from the clinical trials they had conducted or otherwise sponsored. According to the influential *British Medical Journal*, who co-founded the movement, approximately 29% of registered trials had not been reported within 60 months of completion [250]. The AllTrials movement has successfully mobilised substantial public and scientific opinion in favour of their proposition that all clinical trials should be registered and fully reported.

The focus of the AllTrials campaign is on the branded pharmaceutical manufacturer, without differentiating between off-label and on-label trial data. However, shouldn't we be doing far more to address the dangers of off-label prescriptions, where there have been so few clinical trials (and even fewer of regulatory standard) to support the efficacy of the medicines involved? As previously mentioned, not only is there a huge hole related to evidence on efficacy, lacking in three quarters of off-label uses, but also a lack of evidence related to safety, where off-label medicine is between two and three times more dangerous than on-label alternatives. The pharmaceutical industry has responded patchily to the AllTrials campaign, with moves by some companies to increase access to clinical trial data in some areas. Boehringer Ingelheim, GSK, Roche, Sanofi, ViiV Healthcare and Pfizer have all established systems to enable access to detailed trial data including clinical study reports and electronic individual participant data. It is therefore significant that despite these systems having marked differences in approach and scope, they are similar in excluding access to trials concerning off-label uses [251]. Yet, this is the area where the evidence is most questionable!

Let's consider the suggestion that we need more information for all drugs by comparing them with current attitudes to the regulation of driving on the roads. Safety in that regard is an objective that nearly everyone can agree upon; the only question is how to do it. In the United Kingdom in 2010, there were 380 drink-driving deaths; 12% of the total killed on the roads. In other words, 88% of the fatal road traffic accidents were not attributable to drink-driving. In California in 2007, drink-driving contributed to 22% of deaths; or 78% were not caused by drink-driving. Yet to suggest that we should do more to control sober driving before we have clamped down on drink-driving, because this is responsible for the vast majority of incidents, would be seen as perverse, since everyone knows that drinking and driving is dangerous both to the driver and to those they come into contact with.

Are not on-label clinical trials the equivalent of the sober driver, which contribute to 80% of the patient exposure, yet are already held to a much higher standard of safety and efficacy than their off-label counterparts? Should we not be trying to regulate the more 'inebriated' end of the clinical trial landscape? After all, this minority of prescriptions are currently written largely outside regulatory purview. Like the statistics on road traffic accidents, are we not restricting the largely law-abiding driver while the drunk gets off scot-free?

A further problem related to greater transparency of clinical data is that most prescriptions are for generic products. In fact, around 85% of prescriptions are for generic, non-patented medicines. There are many advantages for a system which allows competition to drive down the price after patent expiry, but one of the consequences is lack of accountability, since that competition requires many more than one manufacturer. Figure 8.1 shows the pharmaceutical world in this way, and whether we are likely to learn more about these products and the ways they are used

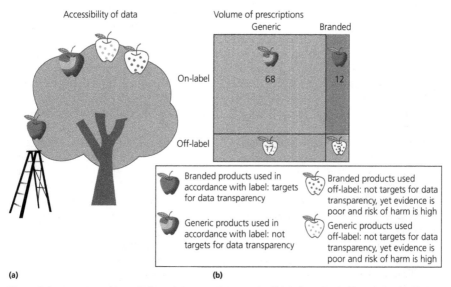

Figure 8.1 Generic and branded products, used on- and off-label, and whether they are likely targets for clinical data transparency **(a)** accessibility of data and **(b)** volume of prescriptions. The areas of the rectangle relate to numbers of prescriptions, normalised to 100%. (*See insert for color representation of the figure.*)

as a result of better clinical data transparency. The depiction of the tree on the left of the figure is meant to represent the parts of the world of pharmaceutical data which are accessible, relative to those that are not – the low-hanging fruit relative to the inaccessible apples. The graphic on the right of the figure shows the proportions of prescriptions which are written for generic versus branded, and off-label versus 'on-label' medicine.

The clinical evidence for the use of most generic products was established many years ago, by organisations that are in many cases no longer holders of a current marketing authorisation; in fact, 45% of generic medicines no longer have an alternative branded version in the market. Let's take a real example. Captopril is a well-known, widely used anti-hypertensive medicine, one of the so-called angiotensin-converting enzyme (ACE) inhibitors. It was developed in 1975 by three researchers at the US drug company Squibb: Miguel Ondetti, Bernard Rubin and David Cushman. It was a time before computers, the Internet and electronic information. Words had different connotations: a web was spun, chips were salted and disks were slipped.

These three and their colleagues built on research by the British pharmacologist John Vane from the 1960s at the Royal College of Surgeons in London. Vane, working with Sérgio Ferreira, discovered a peptide in pit viper (*Bothrops jararaca*) venom which was an 'inhibitor' of angiotensin II, a substance that was known to increase blood pressure. Captopril was developed from this peptide by medicinal chemical research based on the (then) revolutionary concept of structure-based drug design. We now know that ACE is central to the renin–angiotensin–aldosterone system in the control of blood pressure. Captopril was the culmination of efforts by the Squibb laboratories to develop an ACE inhibitor.

It was brand named Capoten™ and gained FDA approval in June 1981. Capoten began an era of markedly improved control of hypertension, but following patent expiry, the product was superseded by generic copies in the United States in February 1996. Nowadays, the originator company does not manufacture any of these products, and instead there are 10 generic companies who manufacture in total 52 captopril-containing products.[1] Captopril remains a hugely important drug for the control of high blood pressure in hypertensive patients, yet its original research, or at least that portion of it that has not already been reported to the regulatory authorities, is largely lost to recorded history.

Squibb is no longer an independent company, having been merged into the much larger Bristol-Myers Squibb. Two of the three original inventors, Ondetti and Cushman, are dead. Even if there are original studies that might conceivably be uncovered through the AllTrials efforts (which I doubt) who would find it, assemble it, check it and make it coherent? If anyone has data, it would be Bristol-Myers Squibb, but the company has no commercial interest in any captopril-containing product. Even if compelled to compile and hand over their entire captopril records, it is not clear that they would have held on to anything other than has already been supplied to the FDA, and that of course was in paper format. Who would convert the reams of documents into electronic format and check the transfers for fidelity? On the other hand, the generic companies, who do have an interest, have no data.

Let me give you another example. Olanzapine is an antipsychotic originally sold by Lilly under the trade name Zyprexa™ but now generic. There are 14 manufacturers listed in as having been approved to sell generic versions by the FDA;[2] none of

[1]Source: FDA Orange Book (http://www.accessdata.fda.gov/scripts/Cder/ob/default.cfm).
[2]Source: FDA Orange Book (http://www.accessdata.fda.gov/scripts/Cder/ob/default.cfm).

them is motivated to spend anything on R&D – even if one of them considered doing so, any benefit from their findings would be distributed evenly among all the other 13 free-riders. Olanzapine belongs to the class of atypical antipsychotics and is intended mainly to control the symptoms of schizophrenia and bipolar disorder, conditions which are very substantially confined to adults. Yet it is used much more widely than that, with the AMA estimating that in paediatric populations, 70–75% of the atypical antipsychotics are used off-label, particularly to treat behaviour problems like ADHD, generally in children, some as young as 2. There is no label for olanzapine to treat this, even for adults, and because the drug is generic, there is very little scope for manufacturers to spend money to evaluate these conditions. We saw in Chapter 1 how the use of these drugs has grown in recent years, along with safety concerns like weight gain, increased levels of cholesterol and higher blood glucose levels leading to an increased risk of type II diabetes. Given these safety concerns, it is particularly important to gather evidence of the efficacy of olanzapine in these off-label indications. Yet, a recent review of off-label atypical antipsychotics found *no* trials of olanzapine in ADHD in children [252]. For advocates of evidence-based medicine, this is completely unacceptable. Even a successful AllTrials campaign is not going to give us much useful information about the egregious use of this drug in this condition; yet it still goes on [253].

I am now going to consider the extent to which the AllTrials proposals address patient harm. To do so, two things are important. First is whether additional data can or will be provided, and second is whether additional data would reduce patient harm. Patient harm depends both on the number of prescriptions and whether those prescriptions are off-label or not. As shown in Figure 8.1, around 85% of prescriptions are for generic drugs. As far as the off-label character is concerned, we have assumed that around 20% of all prescriptions are written for off-label medicine throughout this book. This proportion operates for generics and for branded drugs alike, so we have the four divisions for the purpose of this analysis:

- 'On-label' generics, 68%
- Off-label generic, 17%
- 'On-label' branded, 12%
- Off-label branded, 3%

In total, this adds up to 100%. However, for the reasons explained earlier, the generics (both 'on-label' and off-label) are not generally amenable to further data acquisition. As far as the branded portion is concerned, there have very recently been moves by pharmaceutical companies to provide transparency in their data but only as far as 'on-label' uses are concerned. The off-label information is not formally part of the approved usage for their products, and their concern is that providing this information may expose them to allegations of illegal off-label promotional activity. In all, therefore, around 12% of the prescriptions may be further illuminated by additional information. The vast bulk of prescriptions, 88%, are unlikely to be illuminated.

The second part of the question that I referred to earlier is the association between data and patient harm. Patient harm is greater for off-label medicine, with the relative risk being around two- to threefold relative to the 'on-label' alternatives (see Table 5.1, p. 72). Furthermore, as alluded to all along in this book, there is a lack of data available on the efficacy of off-label medicine, with around 75% of such uses being supported by little or no good scientific evidence [9,43]. So, as shown in Figure 8.1, the group of drugs with the greatest risks and least evidence accumulate at the top of the apple tree: they are the most likely to contain the rotten fruit yet are the least accessible, whereas the AllTrials campaign for data transparency focusses on drugs which are the lowest hanging fruit on the tree, yet should not be a priority for our concern.

The access to information runs in exactly the opposite direction to where it is most likely to cause patient harm: the area of most harm and least data centres on off-label prescriptions while the area of greatest data availability and least harm is focussed on branded drugs used in accordance with their label. We are not going about the problem in the right way; we need to differentiate the desire for more data from the need to prevent patient harm.

This point is perhaps most clearly exemplified by comparing the cases of Tamiflu™ and Levaquin™, the latter being an antibiotic that is often used off-label for the treatment of viral respiratory infections. Tamiflu (oseltamivir) is an antiviral produced by Roche for the treatment of influenza and has been the poster child for the AllTrials campaign. In fact, the other examples used in support of the movement including Vioxx™ and Paxil™ are not issues that suffer mainly from 'hidden' clinical trial data, although they do involve patients being harmed but for other reasons. Information on the cardiovascular safety of Vioxx was made available to regulators and to the public long before the drug was eventually withdrawn from the market; it was just that the wrong comparison was done, and the results were misinterpreted (see p. 54); in the case of Paxil, I see this primarily as an issue of off-label use, since the 'hidden data' were in relation not to the approved adult use, but to the non-approved use in younger patients; this has been dealt with previously on page 84.

Tamiflu has been criticised as being inaccurately portrayed by its manufacturer for an overstated clinical benefit in the treatment of 'flu, whereas in truth it offers insignificant benefits, and exposes patients to the inevitable harms of the drug's side effects. The charge comes from the Cochrane Collaboration which spent nearly 5 years dragging the full set of clinical trial information out of Roche [254]. Yet the analysis contrasts with a wide-scale observational study from Nottingham University, of approximately 30 000 patients during the H1N1 flu pandemic of 2009–2010, according to which early Tamiflu treatment (within 2 days of symptom onset) reduced the risk of death by approximately half, compared to no treatment [255]. There remains a scientific controversy about the Cochrane report, with some 'flu experts favouring the more extensive observational evidence in support of Tamiflu over the smaller scale randomised controlled trials used exclusively by the Cochrane group' [256].

It is possible that both accounts are correct, however, with Tamiflu both able to reduce the most serious effects of 'flu in rare cases, but for this effect to be statistically swamped by the much more common milder cases, in which the drug has little or no overall effect. Whatever the true story, my point is that Tamiflu is really an isolated (and contested) example where the hidden data problem *may* be at issue. By comparison, Levaquin is but one of a class of fluoroquinolone antibiotics which have been far more widely used and, according to the Cochrane group (in another analysis), 'inappropriately' so. In the United States, $1.1 billion was spent annually on unnecessary antibiotic prescriptions for viral respiratory tract infections, compared to the $1.5 billion spent as a total, non-recurring, purchase for the Tamiflu stockpile [134]. Yet the off-label use of fluoroquinolones for upper airway viral infections continues largely unchallenged year upon year, even though the data are not hidden but are clear and public: this form of treatment lacks evidential support since viruses are not killed by antibiotics [131]. A fuller discussion of fluoroquinolones and the patient harm associated with them is to be found on page 76. This is certainly *not* an isolated case: there are many examples of similar situations as described in this book, in all areas of therapy, where off-label medicine lacks evidence and causes significant harm.

Unfortunately, the AllTrials approach will accumulate millions of pages of data on safe and effective drugs because this is easiest to provide, rather than focus on what is needed for better patient healthcare. I would draw the analogy of the drunk man looking for his dropped car keys in the dark car park; in his confused state, he searches in the pool of light shed by the sole lamp post because that is the most illuminated area. Returning by the sober light of day, he realises he has wasted his time: the solution to his problem lay in the unlit shadows.

It's time we got the torch out.

A solution based on increased regulatory supervision

Following the Mediator scandal in France (see p. 102), a wholesale review of the regulation of off-label medicine use was undertaken there, and a new procedure to prevent the reoccurrence of the problem put in place. At its heart was a new French law aimed at strengthening the safety of medicines and healthcare products (Law Number 2011–2012) and a related decree regarding 'Temporary Recommendations for Use' (TRUs; Decree Number 2012-743) [257].

The need for a TRU can be suggested by many competent healthcare institutions as well as patient advocacy groups but may only be authorised if there are no other appropriate medications available for the respected indication. Under a TRU, a 3-year-long observation window is opened in order to assess the benefits and risks of a marketed drug for an unlicensed indication and to collect information on its safety. In that window, the manufacturer can expand its marketing authorisation through the usual procedures, and products are reimbursed according to the new use. Pharmaceutical firms have the responsibility for controlling off-label prescribing, monitoring prescriptions and not marketing outside licensed indications. Further off-label prescribing must be advised to the French Agency of Medicine and Health Product Safety (ANSM).

The TRU involves the marketing approval holder of the drug and the ANSM entering into a formal contract, requiring patient follow-up and reporting to the agency around the safety and efficacy of the new use, with costs borne by the pharmaceutical company. If a risk to public health is found or there is a lack of patient follow-up, the ANSM can modify, suspend or withdraw the TRU. Issuance of a TRU is based on expert advice from a committee, taking into account four factors. These are the available evidence (preferably a controlled trial and rarely observational data), safety (including risks with certain patient groups), the seriousness of the disease and its prevalence.

The French regulations around the further control of off-label use are a welcome attempt to rein in some aspects of the practice, and they may become more widely adopted across the EU, under a framework administered by the EMA. However, they involve a high degree of regulatory involvement and are complex to administer. For each TRU, and one could imagine there might be many for each individual product, an expert committee needs to be established and procedures adopted for patient follow-up and monitoring by the pharmaceutical company. In addition, TRUs may only apply for conditions lacking any licensed products. We will have to see how it works in practice.

I see some weaknesses: including the fact that TRUs are designed to address off-label use in rare and serious conditions, while in practice, prescriptions are written for quite common and non-serious indications. Then, what happens to off-label use of generic medicines, where the manufacturer is not set up to conduct all the monitoring activities? What happens if the required evidence for regulatory approval is not gathered in the 3-year maximum time period for the TRU to operate; is the off-label use then established in medical practice, despite the lack of a formal approval? And finally, how could this work in the United States, where off-label medicine is so commonplace and yet the FDA is prohibited from interfering in medical practice? Some of these disadvantages have been referred to by a French working group [258].

By comparison with the French proposals, some US academics in 2010 suggested an enhancement to the Orphan Drug Act of 1983 to allow off-label use of drugs for rare diseases [259]. Under the plan, companies would be able to promote a product for the new use for a specified period, but would be responsible for monitoring the safety and efficacy. After a notional upper limit of, say, 4000 patients had been treated, a regulatory application would need to be made for the new use. Regulators would be involved in approving the promotional material, analysing the monitoring reports and assessing the data in the new application. One of the problems with this approach is that the normal procedures for approval of a product, requiring prospective controlled trials, would not be followed; instead, regulators would rely on a more detailed form of pharmacovigilance. Other problems include the fact that the scheme only relates to off-label use in rare diseases and that there is no mechanism for regulating off-label use of generic medicines. A variant of this proposal is for new statutory provisions authorising the FDA to mandate safety and effectiveness data for off-label uses that are widespread or based on questionable evidence [260]. It is unclear, however, the exact parameters under which this would operate. Nevertheless, these proposals are constructive, and it is a great pity that, so far, there has been a muted response to the ideas.

My solutions

The solutions I will now adumbrate fall into three categories. These are around:

a Professional standards

b Reimbursement/pricing

c Outcomes

Together, these are conveniently abbreviated as 'PRO' and represent a comprehensive view of how to treat the whole area of off-label medicine. I have considered involving the regulatory system further, but I want to avoid unnecessary encumbrances or partial abolition of off-label prescriptions. The changes I will propose are the minimum necessary for a better system of healthcare, focussed specifically on ways to improve patient outcomes.

Professional standards

The prescriber may be legally unencumbered in the use off-label medicine, but he or she is bound by various professional ethical rules which impinge on the freedom to practise in this way. Most clinicians may perceive off-label prescribing as appropriate and believe that the benefits outweigh the risks. However, their awareness of consequences both to their patients and to themselves appears to be minimal. Clinicians have been described as simultaneously having a generally low level of concern about risk to patients of side effects, unevaluated efficacy and issues surrounding informed consent [44], while at the same time being insufficiently aware of the means to protect themselves from litigation should anything go wrong [261].

We start with the Hippocratic Oath, where the requirement to treat patients with justice leads to the principle of consent, and the requirement for physicians to pass on their learnings as a teacher underpins the need to record outcomes where new knowledge is generated. We shall deal with outcomes later, in trying to fill in the evidence gap by finding ways to measure the safety and efficacy of off-label medicine.

These principles of the Oath are then embodied in the professional guidance from medical associations like the GMC in the United Kingdom and the AMA in the United States. The GMC actually has a sensible set of rules around this, similar to the situation in Germany. These rules differentiate situations in which off-label medicine is the only possible course, relative to situations in which there may be approved as well as unapproved alternatives. Implicit in the UK rules are an obligation for the doctor also to know the regulatory status of the prescription he or she writes. In Germany, the severity of the patient's condition and the availability of licensed alternatives go together to establish a framework around which a doctor can legitimately prescribe off-label. It is clear that this framework does not abide the idea of off-label medicine as routine, but as an exceptional circumstance. In Italy, a more restrictive regime is in place, as we have seen, and there are legal restrictions on the prescription of an off-label medicine that has not been proven in Phase II clinical trials. Yet, rules aside, the gap between the patient's knowledge of off-label medicine and the frequency with which it does in fact occur suggests that the professional rules are not being adhered to as far as patient communication is concerned.

Outside Europe, the professional rules in North America offer much less detail in this area. The AMA says almost nothing specifically about it in their Code of Ethics section [82], requiring the physician to comply with 'good medical practice', though not defining what that is; to provide patients with all 'relevant medical information', although that may be interpreted differently by patient and by doctor; and to obtain 'informed consent'.

Despite the lack of specific guidance from the AMA, it seems axiomatic that good medical practice requires treatment of patients with safe and effective medicines, and that regulatory status is actually a proxy for safety and efficacy. This is particularly true since it appears that most patients believe that regulatory approval codes for safety and efficacy: regulatory status is therefore 'relevant medical information' from the patient's perspective, about which the AMA Code of Ethics requires physicians to inform patients. On this basis, it is reasonable for the US professional bodies to apply similar standards in the area of off-label medicine as do the European bodies, namely, that medical professionals (a) know the regulatory status of the medicines they prescribe, taking note of any regulatory safety warnings; (b) convey this to their patients and obtain their informed consent; and (c) only adopt off-label medicine in exceptional circumstances where they have assessed the evidence and found their prescribing choice to be sufficient for their patient's interests. In my view, this should be the basis for 'good medical practice', and the AMA should say so specifically.

It is notable that in the United Kingdom, guidance on this issue from National Institute for Health and Care Excellence (NICE) says: 'The prescriber should follow relevant professional guidance, taking full responsibility for the decision. The patient (or those with authority to give consent on their behalf) should provide informed consent, which should be documented'. But this is not just a European-centric view of how the prescriber should approach the issue; it is essentially the position of the well-known US bioethicist Alexander Capron [260]. Similar arguments for regulation of physicians' ability to prescribe off-label medicine have also been forcefully made by two US academics, Philip Rosoff and Doriane Lambelet Coleman. In addition, they propose sanctions for violations, varying from financial penalty to suspension of medical licences for practitioners who pursue problematic off-label medicine [262].

In my view, the laxity of professional rules in the United States has provided a conduit for off-label medicine to become a dark art of pharmaceutical sales, and marketing divisions intent on obtaining the maximum commercial gain from their branded products. The lack of controls from the AMA in this area has necessitated the FDA step in with its own sanction of this activity, which of course is necessarily directed towards promotional activity rather than prescribing activity, as it has no mandate to regulate medical practice. However, since the beginning of this century, even the enormous fines levied on the industry have failed to catch up with the commercial advantages of the strategy, a strategy that finds its foundation in the principled desire of the US medical community to prescribe off-label medicine outside regulatory control.

My response to the proposal that US regulators are prohibited from interfering in medical practice is to draw the analogy with off-piste skiing. Imagine, if you will, yourself as a skier intent on descending effectively and safely from the summit to the bottom of a mountain. Fortunately, you have a guide. He can point you down the marked routes which the ski resort administrators have kindly smoothed and cleared of rocks, and which are not overtly steep, or likely to end unexpectedly at a cliff. But the guide is not bound to offer you this 'approved' route, nor is he obliged to tell you when he advises you to pursue an off-piste alternative. His professional responsibility is to 'do no harm', but his legal obligation is not fettered by what the resort administrators deem to approve. Now, this freedom could work in your favour if, for instance, there is a problem with the approved routes on a particular day, or if the mountain is so difficult and rarely skied that there are no approved routes. Then, of course, the expertise of the guide is invaluable. But you would expect the guide to tell you if he had been warned that the off-piste route was prone to avalanches and that if there was an alternative, you would prefer not to be offered the off-piste option. And whether you preferred it or not, you deserve to know what you are likely to have to encounter in the course of descent to the bottom, including whether or not it involves an off-piste route.

Off-label medicine may be good in some circumstances, but the patient should be informed and consent in advance to what is in prospect. When the patient is unable to consent, that prerogative is seconded to a carer, parent or other responsible person. The fact that a treatment is not of approved standard is something that a patient wants to know about (who wouldn't?), and so it is the moral duty of the doctor to inform the patient of this fact. Informed consent is a principle that applies generally across medicine and, apart from emergency situations where the patient is unconscious and there is no time to obtain consent from a relative or other personal representative, should apply in practically every case.

An Australian working party of senior clinicians has proposed a tiered procedure for obtaining consent depending on the situation [44]. The article segregates off-label medicine into three acceptable categories: where the treatment is justified by high-quality evidence; where it is to be used as part a formal investigational trial; or in exceptional circumstances, as justified by the particular individual situation. In the first category, where the medicine is justified by high-quality evidence, the authors suggest that oral informed consent is sufficient. In the second category, formal investigational clinical trial situations mandate informed consent as part of the protocol, in written form; and in the third category of exceptional situations where high-quality evidence is lacking and it is not part of a trial, they also propose that written informed consent is obtained. In this latter situation, they also propose that the treatment is authorised by a local Drug and Therapeutics Committee. Unfortunately, it would seem these sensible proposals are the exception rather than the rule in normal day-to-day medical practice.

Patients are therefore largely misled into thinking the products have received regulatory review, by a physician who either does not know it actually has not been approved for the intended use, or does know but does not care or wish to tell the patient. Regarding the first possibility, objectors to controls on off-label medicine have suggested it to be an onerous burden for a physician to discover the regulatory status of a product. This is not true. In fact, the information about regulatory approval is readily available in the approved product leaflet and well-known books such as the British National Formulary and US Physicians' Desk Reference, as well as numerous online services. In addition, the FDA has taken further steps, requiring drug manufacturers to provide the approved uses and side effects of their prescription drugs in a computer format that is readily accessible to doctors' computers and handheld devices. But regardless of whether it is onerous or not, patients want and/or need to know, and it goes to the safety and efficacy of the prescribed medicine, so it would seem to be part of a medical professional's job. Objections along these lines are therefore specious.

As for the 'don't care' category, we should start with creating a better public awareness on the issue, since the vast majority of patients are unaware that off-label prescriptions can be written, in the mistaken belief that all their medication has been thoroughly reviewed by government-backed regulation. As reported in Chapter 3, 86% of the general public surveyed lacked knowledge about the use of unlicensed medicine in children, an area where the issue is highly prevalent, and there has been much publicity about the issue of paediatric medication failing to obtain formal regulatory clearance. Presumably, there is even poorer awareness about other areas of medicine where off-label use is less common, but still practiced. Even though the public often do not know about this issue, they do care, with another survey showing a level of 81% concerned about the regulatory status of their medicine.

The 'professional standards' part of the solution addresses the ethical issues around off-label medicine. In a way, these are the easiest to deal with. The next two sections try to find ways to reduce the commercial attractiveness of the area, in order to reduce pharmaceutical promotion and consequent patient harms, and to provide incentives for greater evidence gathering and outcomes measurement.

Reimbursement and pricing

Most of the previous proposals have focussed on quite complicated regulatory controls on off-label medicine, without taking advantage of one of the easiest and (so far) unadopted means to deal with the issue. Pricing is a constraint on off-label practice that applies pressure on the manufacturer in a far more flexible and effective fashion than through regulatory action. The principle we can apply is based on the value of the product, which is clearly less in situations where there is less evidence to support efficacy and if there has not been a proper risk–benefit assessment made.

A significant problem of the current arrangement is that the pharmaceutical company who sells the off-label medicine to the healthcare system does so at the same price as the on-label version that they have striven to get accepted through a rigorous regulatory review process (rigorous in order to protect public health): in other words, the off-label product costs the same but lacks the justification normally accorded to pharmaceuticals for their high price, namely, that they have incurred enormous R&D investment in order to get approved. In addition, an off-label product is supported by little or no assurance as to its utility, as we saw in Chapter 6, where we discussed liability issues. Because it is not labelled for the off-label use, the patient's rights to redress from the manufacturer in case it does not work or incurs adverse effects may also be substantially restricted.

In the United Kingdom, through the NICE, we already have restrictions on the reimbursement of products approved since 2001 according to use and patient characteristics. Similar constraints in the United States apply through the reimbursement policies of Medicare and Medicaid. But I want to carry the principle further, with proportionate *pricing* for products according to the level of evidence in support of the safety and efficacy of the product. Importantly, this policy will not interfere with the obligation of the US public healthcare sector to reimburse drugs for off-label use in cancer, merely adjust the price according to the evidence on their efficacy.

At present, the decision is a binary one: either the product is paid for at a fixed price or not. No one has yet put in place a system in which the manufacturer is paid, but at a lower price for a product that is not prescribed in accordance with the conditions of the regulatory approval. Please note that the principle of pricing variation is subtly different from the principle of reimbursement. The latter controls how much an insurer or governmental healthcare body (the payer) reimburses the patient; pricing is an issue between the manufacturer and the payer, and does not affect the patient. Having a variable pricing system has its complexities, but it is not without precedent. Healthcare systems are increasingly intent on observing value from providers, and in some recent cases, pharmaceutical companies have actually agreed to price medicines based on therapeutic outcome. In 2007, for example, Johnson & Johnson agreed to offer rebates to its myeloma drug Velcade™ for patients who failed to respond with reductions of at least 50% in the cancer biomarker within a 6-month treatment period. A pricing regime based around off-label medicine would, by comparison, be much easier to administer because the determinations can be made at the time of prescription, rather than having to wait until a treatment response can be observed, and then there being a certain ambiguity about whether the desired response have been obtained. There are a number of further advantages with this approach.

The first is that with the correct pricing structure, the manufacturer could be highly motivated to react to the real prescribing patterns since the stock of product provided to the healthcare system has a variable value depending on how it is prescribed. We will probably want to put in place a structure such that if prescribed off-label, the manufacturer makes only a small profit, but the actual level of that price needs careful thought. We want to ensure that the manufacturer is disincentivised to promote off-label, and the measure of that disincentive is purely financial, applied immediately at or around the time of the prescription. The current system, remember, aims for litigative penalty many years after the infringement has occurred and depends on proving off-label promotion by the manufacturer. In the case of Xyrem™, this approach foundered in the United States under the free speech defence because the promotion was for a legal act, that of prescribing outside the approved label. In the United Kingdom, the hold the government already has on price has been associated with fewer off-label conflicts and less egregious commercial behaviour, suggesting that the price metric is a good way of encouraging a cooperative approach. With this new system, we don't care how the prescription arose, since we are saying the value to the system of a product backed by poor evidence and with inadequate safety is worth less to our healthcare system.

The second advantage is flexibility and speed. Price is something that can be negotiated between payer and manufacturer much more quickly than regulatory action. Indeed, in the future, it should be an aim to negotiate off-label prices at the same time as the initial pricing negotiations following marketing approval for the main use. The third advantage is probably the biggest of them all; it offers a solution to something everyone in healthcare needs, finding ways to incentivise better evidence for off-label medicine. This advantage, even if all the others melted away, would be more than sufficient to overcome the inevitable disadvantages, which I discuss later. The mechanism by which this would operate is to offer pricing for new uses or patient circumstance at the same level as pricing for the initial use, consequent upon proper evidence being obtained from randomised trials and regulatory approval. Products can then be promoted within the new label.

Let me give you an example of how this could work. Let's take Xyrem (sodium oxybate), which is approved for narcolepsy but used in a wide range of off-label situations. It may well work in those situations, but we don't know how well, nor whether the safety of this street drug is appropriate for these off-label uses. The price, as we know, is high, around $7000 per month. This price is justified on the basis of the indication being for a rare condition, so we can presume the manufacturer has worked out the commercial case on the basis of a very small market size. We also know the manufacturer has been intent on pursuing these more prevalent off-label areas, but has not obtained regulatory approval outside narcolepsy. Finally, we know that the concerns over abuse, and associated side effects of CNS depression and misuse, mean it can only be prescribed in accordance with a restricted distribution programme. Now, say we reduce the price outside narcolepsy to $50 per month, 7% of its 'on-label' price, a level which we assume is commensurate with the costs of goods for this product but also allows for operating the addiction monitoring scheme. Xyrem represents an extreme example of the discount that would be appropriate for off-label uses, as a result of the enormous price for its main, rare application. In other more mainstream products one might presume, on a case-by-case basis, a more modest fraction would be appropriate, perhaps up to 20% or 30% of the 'on-label' price.

However, regardless of the exact figures and ratios, the intention is to incentivise the manufacturer away from off-label sales: to define as clearly as possible how and under what conditions the product really works, to do that as quickly as possible and then to restrict marketing practices to that area. We should remember there is a potential problem with widespread exposure of a product like Xyrem beyond those serious cases that truly warrant treatment with a drug that is accompanied by serious side effects. Once so defined, the use of Xyrem corresponds to the evidence of its safety and efficacy in properly conducted trials; the manufacturer can legally promote in those areas and stand behind any untoward outcome (which, of course, we hope never happens). Flimsy 'marketing' trials can work against the company interests, since they no longer return revenues based on 'on-label' prices, but can create huge liabilities, potentially destroying the main franchise. As far as the prescriber is concerned, no longer is he or she working in a legally exposed grey area; he has the information on how the product works and how well it works; and, moreover, he knows that outside this zone is an area of both unpredictable and unlikely utility.

Why do I think this would work? Well, let me take you back to the story of amiodarone, which was originally available in Europe but was taken up by US physicians for cardiac arrhythmias and made available to patients for free on a compassionate use basis. As soon as the level of the compassionate use increased beyond a certain amount, the manufacturer became concerned about how much this was costing and pressed for US regulatory approval to be granted. The company was able to threaten withdrawal from the programme, so denying access of patients to the drug. But in the case of an off-label use, the product could not be withdrawn without the manufacturer also losing the revenue for the approved use, so the tendency would be to work with regulators to provide the proper evidence as quickly as possible, without the threat of withdrawal of the product hanging over patients.

It would not be fair to fail to mention the disadvantages. To start with, we need to get compliance from the medical profession in terms of formally admitting the fact that a prescription is off-label. We need explicit information on the medical diagnosis and the indication being treated. In fact, this is something inherent in my call for greater transparency in the relationship between doctor and patient, and even something that has been specifically called for in the United States, to ensure that reimbursement is limited to drugs provided for medically accepted indications [263]. We want to avoid 'diagnostic shift' and want to work with professional bodies to ensure this does not happen. A second consideration is risk for the manufacturer, since the value of their stock is dependent on the use to which it is put and therefore indeterminate; but this is no more than the inverse of the risk to the healthcare payer, who is currently required to pay for a product whose value is indeterminate. In other words, it may be used with good effect in accordance with the label some of the time, but off-label with poor effect (and to produce harm, but at the same unit cost) the rest of the time. However, the downside risk to the manufacturer can be rectified once the product is approved for the new use, something that is in their control to bring about; if, on the other hand, it is unworthy for the new use, the manufacturer has a financial incentive to convey this information to the physicians, as a form of negative promotion, in order to limit the frequency with which stock is sold at a lower price in an off-label fashion.

Third, the new system imposes more complexity on an already complex situation; payers will need to put in place payment systems that are robust to avoid abuse (e.g. situations where a prescriber may inappropriately seek to mislabel a prescription for one indication but actually use it for another, to the disadvantage of the manufacturer) and reasonably simple to operate. In practice, this is much more feasible with electronic prescribing systems that are coming into common use – perhaps even contingent upon the advent of such systems. We should recognise that pharmaceuticals are marketed all over the world with different pricing regimes in each country, together with various discount schemes and so on. Therefore, I do not think this represents an impossible burden, nor an unwarranted one. Finally, this scheme can only work with branded products. When genericisation occurs, the prices are flattened, and rather than provide pricing 'sticks' for off-label uses, I propose to offer market exclusivity 'carrots', as I shall now describe.

Outcomes

Off-piste skiing is exhilarating because it involves travelling into the unknown. But when it comes to healthcare, the feeling is rather more scary than exhilarating. If, as an alternative, we had an equivalent where the off-piste runs were marked with the odd flag demarcating the avalanche zone, and other indicators to tell the less adventurous skier where not to go, we would be more likely to arrive at the bottom in one piece. Similarly in off-label medicine, one of the main challenges is that the outcomes are not recorded, or at least not in a fashion that commands scientific respect.

Supporters of medical innovation are keen to point out that off-label practice sometimes meets standard of care. But how do we know, when as a society we regard as 'too risky' products associated with adverse events that occur only once in a thousand prescriptions? Individual doctors are not going to have access to this number of patients. On the issue of efficacy, how can we know this if we rely on underpowered or unblinded or observational trials? And even when the practice is backed by peer-reviewed papers, how can we be sure these are representative when bias in publishing has selected the positive and rejected the negative outcomes?

The way we would like to know this is by making the manufacturer oblige us with undertaking the formal development of the product for the new use or in the new patient group. Some years ago, when the problem of off-label paediatric prescription was noted, a carrot was offered to the industry in the shape of an extra 6 months of commercial exclusivity if such studies were done. In 2002, the US Congress enacted the Best Pharmaceuticals for Children Act, granting 6 additional months of exclusivity (after all other forms of exclusivity have expired) to drug manufacturers who conducted paediatric clinical studies on their marketed product and developed useful information about the safety and effectiveness of their product in children. In addition, the United States enacted the Paediatric Research Equity Act, a stick which allows the FDA to require paediatric studies. The US legislation was a little while in gestation, having been initially proposed by the FDA. Paradoxically, even though clinicians benefitted from the greater availability of data to support their prescribing decisions, the Association of American Physicians and Surgeons, the Competitive Enterprise Institute and the Consumer Alert filed a lawsuit against the paediatric rule in December 2000. The lawsuit, although successful, was eventually overturned by rapid legislative changes that passed through the US Congress a couple of years later.

In Europe, regulations were also crafted in the form of both a stick and a carrot: variations or extensions of existing drug approvals to the EMA must have a compliance-checked Paediatric Investigation Plan that either has paediatric data, a deferral or a waiver for each defined group of paediatric patients at the time the regulatory submission is made. There has also been a move to set up a European clinical trials network in medicines for children, with national networks already established in some EU member states, such as the United Kingdom.

The paediatric regulations have markedly changed the landscape for new drugs, with paediatric developments now more a rule than an exception, spurred by the additional supplementary patent coverage. However, the situation has not changed in the area of generic drugs. Here, the reward is not available, since there is no patent protection to extend. In this category, paediatric studies need to be supported by public funds, from, for instance, the NIH in the United States or EU scientific programmes in Europe, so developments of this kind are still rare [220].

We have another problem too, in that the evidence gathered in support of paediatric medication involves formal approval studies of a prospective nature. There is an essential philosophical difference between using these as a basis for assessment of outcomes, and off-label medicine, as I shall now explain.

Off-label use is not a trial

An off-label prescription, even one backed by no scientific evidence whatsoever (and therefore essentially experimental in nature) is not legally considered a trial. If it were considered a trial, it would require formal protocols, written informed consent, statistical power calculations, adverse event reporting, ethical committees, report writing and so on. The rules around the conduct of pre-marketing clinical trials (so-called Phase I, Phase II or Phase III trials) are fundamentally different from the way in which prescriptions are written. All of this is embedded in the trial conduct, legally constituted under the principles of 'Good Clinical Practice'. However, even though patients would be better protected under such a system than current off-label practice allows, a medical trial involves substantial additional controls that, if they were in place, would transpose it into a substantially impractical option for a prescriber. This is critical to the position of the medical profession in opposing such restrictions around off-label practice.

Regulators govern the conduct of clinical trials, but as we know, the FDA is prohibited from regulation of medical practice. Part of the justification for regulatory oversight of clinical trials is the increased safety risk and decreased evidence base in this area; but is that not also the case for off-label treatment? I have been making exactly the same point throughout this book that this practice is associated with greater risk and impoverished with evidence on efficacy; yet I am hesitant about categorising *all* off-label medicine as a trial, to be surrounded with the same legislative framework as exists for clinical developmental work from which we derive our pharmacopoeia. My concern derives from the impediments this blanket categorisation might impose on the treatment of seriously ill patients where there are with no licensed medicines.

How do we arrive at this somewhat nice distinction? If we go back to the thalidomide example in the first chapter, the difference is down to the intent to treat rather than to evaluate. In that case, the discovery of the anti-leprosy effect arose from a rather time-constrained need to treat a leper on the verge of death. One can see the problems that would ensue if the thalidomide treatment had been delayed while a clinical trial protocol was written, an ethics committee was convened and the drug product obtained (along with a placebo control) for inclusion in a formal trial. The following distinction has therefore been proposed: a drug's use could be classified as a medical practice if the purpose is to 'provide diagnosis, preventative treatment or therapy', whereas it could be defined as research if it is designed to 'test a hypothesis, permit conclusions to be drawn, and thereby to develop or contribute to generalizable knowledge' [214].

Despite the formal difference between the use of off-label medicine in therapy and a trial to prove its hypothetical effect, there is a linear model of innovation that connects the two things. Essentially, what should happen is that a hypothesis should give us an off-label treatment which should give rise to an outcome, and that outcome should inform the next step, be it another off-label treatment or a more formal trial-based assessment. The first time this happens, we are of course willing to accept that a patient in dire need is not best served by the time-consuming procedure required for initiation of a trial. But once off-label medicine is used regularly (and regularly used it is), the standard of care is poorer if evidence from prior cases has not been obtained. It goes against the scientific method to continue to prescribe non-evidence-based therapy: the AMA, for instance, requires that 'physicians must always ensure that medical care is provided only on the basis of scientific evidence' (AMA Code of Ethics Opinion 8.055 – Retainer Practices; Section 2). Such evidence requires clinical trials.

There is also a serious problem when the outcomes from off-label medicine are not written up, even though unfortunately this is more often than not the case. The knowledge of whether the treatment was successful, or (worse) unsuccessful, whether there were safety issues and so on is then retained within the individual medical practice, or hospital, and not passed on externally. Indeed, there are countless examples of off-label medicine which are essentially experimental, the results of which we know almost nothing about. This is a great pity, since as we also know, there are billions of off-label prescription events every year, and if healthcare professionals did write them up, they would inform successive practitioners who might be thinking of doing exactly the same thing. If we were to adopt the language of the AllTrials movement, this is all 'hidden' data, that lies buried in notes from doctors' consulting rooms and hospital wards.

Support for greater dissemination of off-label outcomes actually comes from within the medical profession. A report from the UK Royal College of Psychiatrists in

2009 called, among other things, for the clinician to 'consider writing up the case, to add to knowledge about the drug and its use' [36]. Unfortunately, although this was published in January 2007, its recommendations are still 'under review' and have not been acted upon. There is also scope for publication of individual case reports in the scientific literature. Alexander Capron, the US medical ethicist who was mentioned earlier, believes that a physician who has found an off-label use of a medication to be helpful for a particular problem in a particular population has an ethical obligation to publish a research article reporting the findings to the world. He has even given a threshold outlining when the practice of medicine ends versus when the practice of unethical, non-institutional review board-approved research begins. Once a physician finds a new off-label use helpful in more than two patients, he or she is engaging in unethical conduct if he or she continues treating further patients without setting up a research study. Capron's views have met with resistance from clinicians in some quarters who regard this approach as constraining on their practice [264], but are supported by the American Academy of Pediatrics, who recognise there is an obligation to promote knowledge about off-label uses including 'the systematic development of the information about that drug for the benefit of other patients' [260].

To me, those clinicians who reject the professional responsibility to assist in evidence gathering and dissemination of their own off-label prescribing practice miss the whole point of medicine, which surely is to do the widest possible good to the greatest number of people. It is said that if you give a man a fish, you feed him for 1 day, whereas if you teach a man to fish, you feed his whole family for a lifetime. Hippocrates knew what he was doing when he wrote his Oath, requiring doctors to 'pass on the art of medicine to the next generation': teaching is an essential ethical cornerstone of the profession. Of course, it is easier for a doctor to continue his practice without the burden of passing it on to others. However, his professional duties are not limited in this way. Absent the teachings of the experimental clinician, patients continue to be effectively experimented upon because little is learnt from their therapy, and nothing is learnt outside their place of treatment.

Once the ethical duty of passing on the art of medicine is accepted, the next question is its testing. Clearly, if the hypothesis is right and the proposed therapy is effective, the earlier and more widely we can use it, the better; but on the other hand, if it is wrong, to continue to pursue ineffective or dangerous treatment is a profound medical error. Again, ethics demands that we investigate which of these is right. This presents a fundamental dichotomy, with no middle ground. It comes as no surprise that a particular doctor who has found (he thinks) that his patients do well with the off-label use of a particular therapy will continue to use this treatment, knowing 'intuitively' that it works. Unfortunately, he may also be objectively wrong (as I pointed out in the section on debunking medical myths on p. 101), and the only way to find out which of these two alternatives is actually true requires conducting trials. Capron is right; the next step is setting up a research study.

Write-up

For a single physician with one patient for whom he discovers a new off-label use, it is a daunting prospect to proceed to regulatory approval, unless perhaps he is working in a large pharmaceutical company; and even here, resource constraints are quite likely to put his discovery behind others of greater commercial merit. So, the process towards such regulatory approval needs to occur stepwise.

Clinic-based discoveries typically start with individual case reports, then move on to case series, an open label trial and then to Phase I/II trials and perhaps beyond. When Dr Sheskin treated his leprous patient with thalidomide, it was the start of a long process of successive trials and ultimately development undertaken by the biotech company Celgene, but as a result of that effort, we ended up with thalidomide being used successfully to treat leprosy and the drug being further developed for myeloma. This is not a solitary example: individual doctors are commonly involved in discovering new uses for existing drugs. A survey of new therapeutic uses for new molecular entity drugs approved in 1998 revealed 143 examples of which 57% were discovered by practising clinicians independent either of pharmaceutical company or university research [265]. This figure demonstrates the special position that clinicians occupy in being able to identify the widest therapeutic application for the armoury of available pharmaceuticals and suggests further that there are many more applications we do not currently know about.

An obvious first place to record such discoveries is as a case report in a peer-reviewed journal. However, discoveries such as these from practising clinicians have normally not followed the standard research development pathway ('bench to bedside') typical to academic or industrial medicinal research. They therefore often lack the mechanistic underpinning and ancillary experiments that good, high-impact journals like to see. While there may ultimately be certain less impactful publications that will accept such case reports, we should take note of the difficulty authors will find (and also note that for the top peer-reviewed journals such as the *New England Journal of Medicine* or *Nature*, the acceptance rate is as low as 6%).

In addition to journal articles, there are other places where the art of off-label medicine, as Hippocrates would have said, can be 'taught' to successive generations. Increasingly, there is interest in being able to capture information from doctors' notes; in February 2014, it was announced that data in the United Kingdom from individuals' GP records would be shared with researchers inside the NHS, thereby making it more possible to link up individual case reports of this kind [266]. A second purpose of retrieving data such as this is not just to provide evidence of therapeutic efficacy, but also to evidence of safety. In the United States, avenues for further information-gathering involve data from electronic health record systems relating to drug prescriptions as an early alert of increasing off-label use of a particular drug [267]. In this paper, the authors a conducted a large-scale characterisation of off-label usage using fully automated textual analysis of physicians' clinical notes, in order to identify potential off-label medicinal uses. They then coupled this with automated searching of scientific databases to provide a ranking of the risks associated with the new uses. A similar approach, though less complex, was used by the healthcare company Medco in an analysis of electronic medical records of patients in their database, with the aim of identifying the extent and variety of off-label use [268].

Electronic healthcare records have enormous potential to offer better understanding of off-label medicine. Their power is built on numbers rather than the individual case histories, on whole population statistics rather than granularity at the patient level. These approaches are still being developed, but they offer a route, even absent the active involvement from doctors, to clarify the frequency of events and their outcomes. This knowledge can then stimulate the manufacturer to properly evaluate the situation.

Pharmacovigilance

We are fortunate when developing additional uses for existing products, or the same uses in different patient populations, in being able to rely on large and growing safety databases for the basic product. This comes about because once approved by the regulatory authorities, medicinal safety is then subject to ongoing assessment through pharmacovigilance programmes, and this process is also managed by the regulatory agencies.

Pharmacovigilance (or post-approval surveillance) is important because rare side effects are usually not evident from the limited number of patients studied in the pre-marketing trials. In order for all the risks we care about to be assessed before regulatory approval, it would require enormous trials which are incompatible with an efficient process for product innovation, as we discussed earlier. Pharmacovigilance programmes have revealed things we could not have known from smaller sample sizes, such as the rare but potentially serious adverse occurrences of sepsis (and even death) associated with anti-tumour necrosis factor therapy (like infliximab or Enbrel for rheumatoid arthritis), side effects which manifest after prolonged exposure (such as hepatotoxicity with low-dose weekly methotrexate for rheumatoid arthritis) or safety events which occur after a long latent period (such as infertility after chemotherapy for cancer in childhood).

In 2012, European pharmacovigilance legislation strengthened its requirement that the marketing authorisation holder should be responsible for continuously monitoring the safety of its products, specifically including the obligation to inform the authorities of *any* use of the medicinal product which is not in accordance with the label. These responsibilities also include reporting on the results of clinical trials or other studies. When a marketing authorisation is renewed, all relevant information on the safety of the medicinal product should be considered. The key features of the risk management system should be included in the marketing authorisation application. In the United States, the FDA has started to build the Sentinel System, which is designed to supplement the post-approval drug safety process by the rapid collection of information on safety and the real-world assessment of clinical outcomes across a range of healthcare environments.

In both the EU and the United States, pharmacovigilance regulations have stopped short of requiring efficacy measurements for off-label products. However, it is worth pointing out that both French and US proposals referred to earlier (see p. 152) involve specific aspects of enhanced monitoring or pharmacovigilance, either as a duty by itself, or in return for an ability to promote in the off-label area [258,259]. Even without these proposals, the safety-monitoring environment is becoming significantly more extensive, particularly in the EU. The 'Black Triangle' system established in the United Kingdom is in the process of being extended across the entire EU and as applied to newly registered products will radically increase reporting since it is compulsory for doctors to report any adverse event, however minor; under the voluntary arrangements that apply for older products, it is reckoned that up to 90% of adverse events fail to be reported. Taken together, it would be appropriate in my view to make the reporting of any adverse event for any off-label medicine a compulsory requirement. This should apply beyond the present 2-year limit after registration which applies for uses within the regulatory label, since many of the situations which involve off-label use are being developed during the course of the lifetime of the product. Doctors are sometimes wary of reporting adverse events resulting from off-label use because of potential litigation. Exposure to litigation may contribute to the low rate of voluntary reporting already noted. The EU pharmacovigilance reform, therefore, aims to tackle this problem by encouraging practitioners to report all problems in a blame-free environment, under a cloak of confidentiality, even in the case of medical errors [269].

Commercial development

With the early evidence of efficacy in the new therapeutic use, or new patient population, combined with the safety information that derives both from the original approval of the drug and increased knowledge from pharmacovigilance, we have a

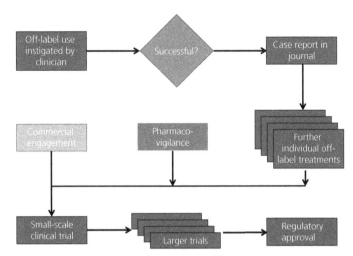

Figure 8.2 Possible development route to convert an off-label medicine into one with regulatory approval. Additional involvement of pharmacovigilance data comes from ongoing regulatory assessment of the product according to the original approval. Commercial engagement is incentivised either as a branded medicine by pricing, or if a generic medicine by exclusivity arrangements.

platform from which to embark on the regulatory development of the off-label use, converting it to approved status. A scheme outlining the main components of this development, starting from the original individual clinical treatment, leading right through to approval, is shown in Figure 8.2.

The depicted outline shows how this can be taken through to the regulatory approval by a pharmaceutical company. One possibility, but not the only one, would see the developing company be the originator who developed the product in the first place. We saw in Chapter 2 examples like finasteride which was developed through to regulatory approval for both prostate enlargement and secondarily for baldness by the same company. In this case, Merck filed a patent for the treatment of male pattern baldness using finasteride, which ran from 1996 until 2013. In comparison, the original composition of matter patent on finasteride expired in 2006, thereby giving an extended period of monopoly for the hair loss product (see Figure 8.3). There was therefore a very clear commercial case for patent-protected commercialisation of the product in both uses. There are other examples where secondary use patents were not necessary to support secondary indications; I am thinking, for instance, of the use of sildenafil (first used for erectile dysfunction) in pulmonary arterial hypertension. In this case, Pfizer commercialised the new product Revatio™ within the time window of the original composition of matter patent for sildenafil. This patent also covered the product Viagra™.

Commercial incentives may also apply for a second company (i.e. not the originator) to develop the new use, provided there is a differentiating factor relative to the product from the first company. The development of doxycycline for periodontitis (gum disease) was carried out by a small company called Collagenex, whereas the original antibiotic product containing doxycycline was developed by Pfizer. The periodontitis product was deliberately formulated to contain sub-antibacterial doses of doxycycline, and it was then launched on the market after Pfizer's basic doxycycline patent had lapsed. Generic versions of doxycycline were on the market by then, but

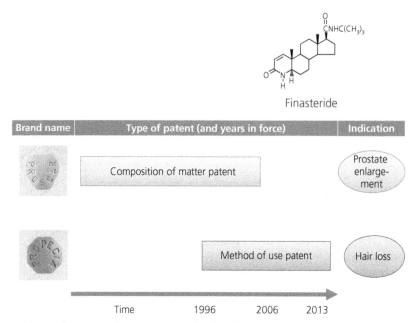

Figure 8.3 Merck's two products containing the drug finasteride were for prostate enlargement and hair loss. The second product (Propecia) was developed under a method of use patent and contained a different dose of finasteride than was contained in Proscar.

they were used for the treatment of bacterial infections and contained too much doxycycline to be safe in periodontitis. This is because periodontitis requires chronic treatment, and exposure to antibiotics over a long period is undesirable for a condition where resistant bacteria can develop.

The opportunities that a second company may identify around a differentiated product, similar to the strategy adopted by Collagenex, are fairly general and apply even after the original product has become generic. Genericisation substantially diminishes the commercial potential for the generation of data on secondary uses, though opportunities can still exist. The key component is having a factor that permits differentiation of the product for the second use relative to the generic and finding a means to obtain a patent to protect the second product from generic competition.

These possibilities are fine when the off-label instance involves a secondary therapeutic use, because it is possible to file patents for secondary medical uses. If one wants to extend the data package for a different patient population, patent coverage is less likely, unless some extraordinary difference is found between the two patient groups, such as an unexpected dose requirement. Having said that, GSK found that apparently there was an unexpected difference in the response to their irritable bowel syndrome drug, alosetron, in women, and managed to obtain a patent specifically for this treatment group (see p. 79); however, this is the exception rather than the rule. The difficulty in obtaining a new patent for paediatric patients is one reason why the paediatric legislation is framed the way it is, focussed on an incentive ('carrot') that applies to the originator, on trying to get the development work carried out within the time envelope of the original patent period for the product and then extending that exclusivity period by 6 months.

This discussion about patents and commercialisation may seem rather disconnected from patient health; but actually it is vitally important to gain the attention of a commercial partner in order to conduct the development and acquire the data needed to substantiate safety and efficacy to regulatory standards, as shown in Figure 8.2. At the end of the day, the drug industry will not produce better evidence unless it is in their financial interest. This point is emphasised by considering what happens once we are in the era of generic competition without any commercial development route.

Non-commercial development

I hope by now I have persuaded you of the importance for healthcare of defining the envelope of therapeutic possibilities for a drug, so that it is used as widely as possible, and with as much evidence as possible. If we are not able to do that within a commercial framework, we are faced with very long development timelines, the kind of problem demonstrated by aspirin.

As discussed in Chapter 2, aspirin has three uses: for the treatment of pain, for the prevention of strokes and heart attacks, and for the prevention of cancer. This last use is perhaps the most remarkable: a drug so commonplace and so familiar to us as a treatment for headaches and minor trauma yet also useful to prevent the most feared of diseases. Of course, its use in cancer is still 'off-label', the formal proof of its safety relative to efficacy not having been established. Aspirin also causes gastropathy (like all non-steroidal anti-inflammatory drugs), so we need to be sure that the anti-cancer benefit outweighs the hazard, and that analysis has not been done yet.

Evidence of a protective role of aspirin on the risk of colorectal and other common cancers has been building up since the end of the 1980s, but it took until 2010 for a large meta-analysis of the effect of daily treatment with aspirin on the incidence of several types of cancer in several trials [65]. These trials were conducted to look at the effect of aspirin on cardiovascular disease, so the retrospective nature of this analysis means it does not conform to the quality needed for regulatory submission. The fact that aspirin is available in high-street chemists, over the counter and very cheaply, means that the costs of doing the work to define exactly what is the best dose of the drug to offer the best profile of cancer benefit relative to gastric ulcer side effects are unlikely to be returned to any company that invests to sponsor this work.

Imagine if at some point in the future, there was an aspirin product for the treatment of cancer. Then, should you be prescribed such a product, your doctor would probably suggest to you that rather than purchase the anti-cancer brand, you could buy the generic alternative sold for the treatment of headaches, which would be much cheaper. This is one occasion when the patient would be able to exercise his own off-label choice.

Therefore, all of the research work conducted on the anti-cancer effect of aspirin has been funded by public sources. Denied a commercial backer, this work has taken many decades and will continue to take a good deal longer. In the meantime, the true picture of the preventative effect of aspirin on many of the cancers we are likely to face is not available to us. Instead, commercial companies invest in new products which they can sell at much higher prices but under the umbrella of patent exclusivity. The principle of genericisation, which has delivered us from expensive branded medicines after patent expiry, and saved our healthcare systems' enormous amounts of money, has also resulted in a commercial framework under which the full benefits of those generic products are denied to us.

Aspirin is not a singular representative of this situation; plenty of generic products have similar benefits which we have insufficient knowledge about. Metformin is another example: it is the most prescribed anti-diabetic drug in the world but has shown additional clinical benefit in many cancers ranging from solid to haematological malignancies. Most of the research on metformin has looked at cancer incidence retrospectively, and a substantial body of evidence has built up supporting its protective effect [270]; however, in addition, there has also been work showing prospectively that metformin reduces the incidence of colorectal, liver and pancreatic cancer after initiation of treatment in diabetics [271]. Despite the encouraging evidence for this cheap and safe drug, its use in cancer remains off-label, and it is uncertain when, or indeed if, we shall ever have regulatory standard proof for a metformin-based anti-cancer product.

In my work to discover and develop secondary uses for existing compounds, the main single reason for failing to pursue an interesting idea is lack of commercial incentive. There are many, many developments that could produce cheap, safe and hopefully effective drugs, but there is little prospect of a return for the person or company who invests in the R&D. This is a calamity for the patient.

One way of improving this situation is to involve non-commercial entities in developments of this kind. In the United Kingdom, the National Institute for Health Research (NIHR) has an annual budget of £1.3 billion of government funds for, among other things, clinical investigations of this kind. The aim of the organisation is to deliver improvements in health which can be of benefit to the UK NHS. Their work includes a specific focus on clinical development of secondary uses for existing drugs, normally beyond the stage where initial positive outcomes have been reported. An example of the kind of work they do includes the development of the antibiotic minocycline for the treatment of certain specific aspects of schizophrenia, such as apathy and social withdrawal, which are poorly treated with current agents. The NIHR-sponsored mid-stage study looked at the effect of adding the drug to the existing treatment regime in over 140 patients and found a significant difference in the minocycline-treated group [272].

There are signs that charities too are realising the potential of using drugs from the existing pharmacopoeia as substrates for new medicines, based on secondary uses. Part of the reason for this is that their funds can deliver much more advanced treatments, with a real potential to treat the people suffering from the disease the charity represents, compared to 'blue-sky' research. There are a few examples where this is starting to occur, such as the embarkation upon a major clinical trial to test the efficacy of the anti-diabetic drug liraglutide (Victoza™) in Alzheimer's disease, funded by the Alzheimer's Society United Kingdom; another example involves support from the Michael J Fox Foundation for a number of clinical trials of existing agents in Parkinson's disease. Among the projects are two more anti-diabetic drugs, exenatide and pioglitazone, which work by different mechanisms.

Cures Within Reach, which is a charity engaged in funding developments for secondary uses of existing drugs in rare diseases, has proposed the use of Social Impact Bonds (SIBs) to provide the resources for these kinds of developments. SIBs, also known as 'Pay for Success Bonds' or 'Social Benefit Bonds', are contracts with the public sector in which a commitment is made to pay for improved social outcomes that result in public sector savings. The first such bond was implemented in the United Kingdom in a non-health field, for improvements to recidivism rates for prisoners; since then, others have included bonds (or discussions thereof) for homelessness and childcare. There are potential problems with using this type of scheme for drug development programmes that are essentially global in impact; by comparison, the improvement in prisoner behaviour is a local and national issue. Local or national governments are unlikely to want to see foreign or neighbouring jurisdictions benefitting from their investment without a contribution having been made. The second problem with SIBs is that healthcare is not always a public sector issue, in particular in the United States, where much healthcare is paid for through private insurance programmes. Nevertheless, this type of approach deserves to be looked at in detail and perhaps adapted in order to allow evidence on off-label uses of generic drugs to be acquired, and potentially regulatory labels to be adapted.

We should remember that there are significant challenges ahead, since the non-commercial nature of the development may need to continue until the regulatory status is approved or at least until sufficient data are acquired to give clinicians the best evidence they need to prescribe. Nevertheless, the initiatives in Parkinson's and Alzheimer's diseases are particularly positive developments, especially as the research and development pipelines of the pharmaceutical industry into these neurodegenerative diseases have been so unproductive in recent years, while prevalence and medical need are already high and increasing rapidly.

If this strategy is successful, should we take it further and involve public money more generally in drug development? After all, healthcare innovation normally starts in the public sector, with an academic discovery, and ends in the public sector, with publicly funded healthcare delivery. As mentioned earlier, in the United Kingdom, the NIHR is involved in clinical developments with potential relevance to the NHS. Indeed, even in the United States, the enormous amounts of money spent on Medicare and Medicaid suggest that it, too, has a substantial publicly funded component to healthcare payment. Between these two publicly funded bridgeheads, the pharmaceutical industry operates a commercial model.

Some would argue, given this situation, why does the state not also routinely fund drug development? There are three answers why this is not a terribly good idea. The first is that drug development is extraordinarily risky, and public governance does not sit well with risk. In democratic systems, there is a reluctance to see tax-payer funds spent on unsuccessful development projects. Seen in retrospect, the abandoned development projects (which, history tells us, would constitute, say, 90% of all development projects) would be huge millstones around public officials' necks; these are consequently a cautious lot. The second reason is that drug development is a global business, yet individual governments would need to find ways of monetising their national investment without allowing foreign governments to benefit from that investment. And the third argument is that where this was tried, namely, in the Eastern bloc countries during the Cold War, the success rate was poor.

Nevertheless, secondary uses for existing drugs represent a lower-risk strategy for pharmaceutical innovation, and there are moves from the US National Institutes of Health to be more active in drug developments of this kind, particularly through its latest new institute, the National Center for Advancing Translational Sciences (NCATS). Part of the rationale for this involves the idea of returning value to society, based on the fact that the NIH is funded overall with $30 billion of US taxpayers' money. NCATS mission is to improve the ways that basic science is transformed into medicine, and it has been particularly active in projects involving secondary uses for existing drugs. With an annual budget of $0.5 billion, NCATS has initiated a number of developments of existing drugs for secondary uses, though so far it has chosen to focus on stalled drugs from failed pharmaceutical company development programmes rather than marketed entities.

These, and other initiatives in non-commercial drug development to support off-label use, are welcome insofar as they deliver good data which clinicians can use to better treat their patients. But I think we can do more with the way prescriptions are filled to incentivise generic producers to take on the burden of additional studies to support the off-label use of their products.

At present, in the United Kingdom, generic drugs are automatically substituted by the pharmacist at the point of filling the prescription, unless the physician specifies that a generic must not be substituted. There are different rules in other countries, but all healthcare markets are becoming more filled with generic products, and both governmental as well as private payers are interested in increasing the scope of generic substitution for costly branded products after their patent expiry. As patents continue to expire year by year, there are a growing number of generic products. As we know, this applies both for on- and for off-label prescriptions, so that generics constitute around 85% of the prescriptions written by doctors in the United Kingdom. That figure is lower in other countries but has been increasing rapidly in recent years; as branded medicine becomes ever more expensive, the incentive to use generics wherever possible increases, and this has been manifest in the United States, rising from 57% in 2004 to 75% in 2009 to nearly 80% in 2011. The pharmacist can choose from an array of generic manufacturers who compete with one another, mainly on price. In order to ensure substitutability, generic products take the existing label of the innovator drug company and copy it, exactly. In other words, whatever label existed for the product at the point the product became generic is frozen. They then reference the information produced by the innovator product and submitted to the regulator. The generic provider is entirely uninterested in developing support for paediatric prescriptions, or for providing evidence for any of the off-label uses of the product which existed before genericisation occurred. As I have said previously, the off-label use of prescriptions before genericisation becomes indelibly imprinted, with no commercial incentive in place to change that situation.

I want to put in place a system to enhance the available data for off-label generic prescriptions and have given this matter a great deal of thought. My proposal is simple: we should permit generic companies to provide these data, obtain regulatory approval and in return reward the first such company to obtain such approval with a period of market exclusivity for those prescriptions covered by the enlarged label but at a price of around 40% of that charged for the medicine before genericisation occurred, adjusted for inflation. In other words, generic companies should sit in the developmental scheme outlined in Figure 8.2 in exactly the same place as the commercial engagement envisaged by branded pharmaceutical companies. Their incentive is a higher price than achievable for other generic companies, but lower than achieved by the originator company, along with a period of exclusivity during which they are the monopoly manufacturers of treatments for the off-label use. There is a parallel to be drawn here with the market exclusivity offered for developments in orphan diseases, in which a period of exclusivity exists for 7 years in the United States and 10 years in Europe. The orphan legislation has been very successful in incentivising drug developments for rare diseases, although there have been some problems. One such problem is that products approved in this area tend to command high prices, as with the earlier example of Siklos. This problem is not apparent with the reference point of 40% of the pre-genericisation price specified earlier, and given which, I would suggest an exclusivity period similar to that operative for orphan products, namely, 7–10 years would be appropriate. However, this is something for consultation and discussion.

Let me give an example. Statins have a multitude of biological effects, with potential clinical applications beyond their well-known ability to produce lipid-lowering in patients with high cholesterol levels. Over the years, many clinical studies have been carried out to look at additional uses of these agents. The majority of these are retrospective associative analyses, some are individual case reports and some are even early prospective clinical evaluations. A few examples of these studies are shown in

Table 8.2 Potential additional uses for statins supported
by evidence in human situations, from retrospective trials,
case reports or prospective studies.

Therapeutic indication	References
Asthma	273
Cataracts	274
COPD	275
Depression	276
Diabetic macular oedema	277
Epilepsy	278
Glaucoma	279, 280
Oesophageal cancer	281
Pneumonia	282
Polycystic ovarian syndrome	283
Prostate cancer	284–286
Psoriasis	287
Raynaud's phenomenon/systemic sclerosis	288
Rheumatoid arthritis	289–291
Sepsis, burn injury	292–296
Transplant rejection	297
Vitiligo	298

Table 8.2, which is not meant to be a complete analysis. If, randomly, we take one of these potential indications, rheumatoid arthritis, there are two associative studies indicating a reduced hazard of rheumatoid arthritis associated with statin use and one later prospective study evaluating the effect of 12 weeks of treatment with ator-vastatin Lipitor™ in 55 rheumatoid arthritis patients. In the latter study, various biomarkers of the condition and the immune function of the patients were assessed. The positive conclusion from the trial was also that the drug reduced disease activity, but this would need to be repeated on a wider scale in order to prove the effect sufficient for regulatory approval – an approval that, despite the early promise, cannot be guaranteed. In the meantime, all the statins are of generic status, with some costing as little as £1 a month, so investment is not a commercially attractive proposition, despite it representing a lower-risk programme than a new discovery project. Nevertheless, it is something that could be developed according to my proposal. By offering a period of market exclusivity for the company who undertook this work, patients could benefit from a new approved product for rheumatoid arthritis that we already know to be safe and well tolerated from its prior history as a cholesterol-lowering agent but still be confident that the product would be available at a reasonable price.

According to this proposal, all generic manufacturers would retain the right to supply product for the originally labelled indications, at a price similar to one another; but the one company who conducted the regulatory trials and obtained approval for what was previously the off-label use(s) would have exclusive right to supply product for those uses, at a higher price. The commercial incentives I have outlined earlier may not work for all drugs in all secondary indications. In practice, the relative attraction of the commercial opportunity compared to the investment required to conduct the formal regulatory work depends on at least three factors: (i) how much data are there already, (ii) how large is the market and (iii) how risky is the new work – for instance, development in some indications requires much longer trials and more uncertain outcomes than others. There may be some situations where higher prices or longer periods of commercial exclusivity are necessary. However, even if the rewards as currently proposed are not sufficient to stimulate all commercial developments, it is my intention at least to stimulate a significant proportion of the opportunities in these areas.

To be fair, the proposal has some difficulties, which I think can be managed, similar to those for the price variance proposed for the off-label use of branded medicines. For one, there is the potential for abuse at the point of prescription; there is also an additional complexity for the pharmacist who needs to adapt to two tiers of 'generic' medicine; and there is complexity within the payment process. In my view, administrative arrangements to cope with these factors are feasible under a suitably configured electronic healthcare record system. The patient's age, the therapeutic indication and the dose govern whether the generic can be substituted at the point where the pharmacy dispenses the product with an alternative from another company or whether market exclusivity applies in favour of a particular manufacturer. It is also the case that not all attempted developments will work: sometimes, a company will embark on a process towards regulatory approval of an off-label use of a product, falsely believing the years of prior use have led to a good deal of medical assurance of the validity of that off-label use. But even when the attempted regulatory clearance fails, medical science will have advanced, the negative outcome will have improved our knowledge on how to use the product properly and warned against continuing inappropriate off-label use. Absent an incentive to conduct these trials in the first place, we would be none the wiser.

I am convinced of the importance of this measure: in fact, I see it as essential to correct what I see as a failed market. It is undeniable that pharmaceutical companies find no commercial gain in properly developing generics for their off-label uses. Yet, most prescription drugs are generic, and most drugs have secondary uses: the undeveloped healthcare potential here is actually very large, not just in the uses which have already been proposed and partially evidenced, but also in uses which are still unknown. The number of undeveloped secondary uses for generic drugs is only going to increase as branded products lose their patent protection, so the problem, without measures to deal with it, grows inexorably with time. Moreover, in order to pursue a commercially profitable course, innovator companies seek to develop new drugs rather than repurpose old ones: rather than develop metformin, for instance (which is safe, cheap and used widely for type II diabetes), for its anti-cancer properties, companies prefer to focus on new alternatives for the treatment of cancer, which are very expensive. The commercial advantages of slower genericisation and higher prices are also behind the amazing swing towards biological products, which comprised 71% of revenue from the top ten products worldwide in 2012, relative to only 7% in 2001 [299]. Yet, these drugs are comparatively poor at addressing CNS diseases and require patients to inject rather than take oral medicines [300]. I do not criticise the pursuit of profit here, merely observe it to be true.

On order to correct this behaviour, we have to regulate the industry more intelligently. If we can do so, the opportunities for the healthcare payer are the following: first, off-label generic medicines become properly supported by additional evidence and formal assessment, from which improved healthcare derives; and second, the payer is either partly or fully protected from the distortional effect of the enormously high costs of some new drugs. I would point out that the prices of new medicines these days are becoming eye-wateringly high, with figures in the $50 000–100 000 a year range increasingly common. Pharmaceutical price inflation at this level is unsustainable (as shown earlier with the example of Sovaldi™). Healthcare systems increasingly work within a limited expenditure envelope, so that high costs for one medicine impose restrictions in other areas. In a fully working system according to my proposals, there is also the potential to uncover and then develop additional uses for generic drugs which are currently undiscovered. Getting to this fully working system is difficult and complicated. The proposals outlined herein are a first step, but implementing them will take time, the engagement of multiple parties and the setting aside of partisan interests for something that is common to us all, our own health: in the end, we are all patients above everything else. It is a journey that, for all its difficulties, is both necessary and worthwhile.

We have now reached the end of the proposed solutions for justifying unapproved medicine. The aim has been to find ways of improving the equity for patients around off-label medicine, remembering that this is an area of high safety risk and poor evidence base. We started with a situation where the pharmaceutical industry, the medical profession and governments are, in some circumstances, looking out for their own interests above those of patients. All three groups may be criticised for hypocrisy: the pharmaceutical industry claim adherence to safety and efficacy mandated by regulatory rules, yet have promoted unapproved medicines in circumvention of those rules and pursuit of commercial profit. Clinicians call for the best evidence possible for the products they prescribe, yet still want complete freedom to prescribe off-label medicine devoid of, or deficient in evidence; and then disregard any obligation to help collect and publicise the outcomes from these treatment options to support their colleagues in similar future positions. And finally governments, while in the guise of regulator, impose restrictions on off-label marketing, but in the guise of payer seek to promote off-label use by subverting the very same regulatory systems.

Conclusion

I therefore argue for increased controls on the prescription of off-label medicine and incentives for its proper evaluation that should include:

Professional rules:

- Clinicians must know the evidence in support of the prescribed medication including the fact of whether the prescription is on- or off-label.
- They should be aware of any clinical guidance that permits or recommends the off-label use, as well as guidance recommending against use in specific circumstances.
- They should inform the patient of the regulatory status of the medicine being prescribed, as well as any licensed alternatives and the risks and benefits thereof.
- They should inform the patient and pharmacist of the diagnosis, preferably by writing this on the prescription.
- Off-label prescriptions should only be written for serious conditions and where a licensed alternative is not available.
- Professional societies should discipline offenders in regard to any of these rules to redress patient harm by drugs that are prescribed off-label.

Reimbursement/pricing:

- Healthcare payers should negotiate reduced prices for branded pharmaceutical products that are sold for off-label purposes; for new products, pricing negotiations for off-label situations should take place concurrently with pricing negotiations for the approved label product.

- The exact price should be targeted so that manufacturers do not profit significantly from these non-approved situations and are instead to be incentivised to provide development outcomes data in these areas.
- Once approved for the off-label situation, prices similar to those for the on-label use are appropriate.

Outcomes:

- In order to encourage evidence gathering on outcomes of use of off-label generic medicines, a period of market exclusivity should operate for generic companies who obtain regulatory approval for their products in areas of off-label use, implemented by the pharmacist or healthcare payer by restrictions on substitution. Products that comply with the expanded label should be sold at, for example, 40% of the pre-genericisation price applicable for the on-label product. Alternative percentages and periods of exclusivity may be applicable on a case-by-case basis.
- Pharmacovigilance: formalised frameworks for compulsory reporting of adverse effects (similar to the 'Black Triangle' scheme currently used for new medications) should apply for all off-label medicine throughout their product lifetime.

The intention of these reforms is not to prevent off-label use, but to reserve it for the rare, exceptional case where nothing else fits the patient's needs and to discourage off-label marketing by pharmaceutical companies. Exceptions aside, this rather grey area of medicine should be brought into line with other forms of modern medical practice, substantially evidence-based and judged to be appropriately safe and effective for the particular circumstances in which it is used. We have known that drugs are often useful for more than one purpose ever since aspirin, which was first marketed at the end of the nineteenth century: new opportunities will continue to flow from the rich vein of over 5000 existing pharmaceuticals, including those we first discover in this century too.

The possibilities open to secondary uses are not fully identified within the 20-year patent life of a typical branded product, yet once genericisation occurs, the commercial incentives become much more circumscribed. So, as time marches on, even though more drugs are open to repurposing in more possible ways, the mismatch also grows between what new uses can be treated and what are commercially attractive to support with good evidence.

Partly, the challenge ahead is fully to explore the utility of our pharmacopoeial armoury. But it is also to ensure that drugs with a secondary use are not administered as second-class medicine. It is surely not appropriate (apart from in exceptional circumstances) to apply these products with anything other than the same ethical standards as were required for their first-approved format.

Only once we have adopted this principle, and put it into practice, can we claim to have justified the unapproved medicine that is currently used so widely in our healthcare systems.

References

1 Minar EL, Edelstein L. The Hippocratic oath: Text, translation and interpretation. Am J Philol. 1945;66(1):105.

2 Lindsley CW. The top prescription drugs of 2011 in the United States: Antipsychotics and antidepressants once again lead CNS therapeutics. ACS Chem Neurosci. 2012;3(8):630–1.

3 Boos J. Off label use – Label off use? Ann Oncol. 2003;14(1):1–5.

4 Cuzzolin L, Atzei A, Fanos V. Off-label and unlicensed prescribing for newborns and children in different settings: A review of the literature and a consideration about drug safety. Expert Opin Drug Saf. 2006;5(5):703–18.

5 Pandolfini C, Bonati M. A literature review on off-label drug use in children. Eur J Pediatr. 2005;164(9):552–8.

6 't Jong GW, Eland IA, Sturkenboom MCJM, et al. Unlicensed and off-label prescription of respiratory drugs to children. Eur Respir J. 2004;23(2):310–3.

7 Knopf H, Wolf I-K, Sarganas G, et al. Off-label medicine use in children and adolescents: Results of a population-based study in Germany. BMC Public Health. 2013;13(1):631.

8 Conti RM, Bernstein AC, Villaflor VM, et al. Prevalence of off-label use and spending in 2010 among patent-protected chemotherapies in a population-based cohort of medical oncologists. J Clin Oncol Off J Am Soc Clin Oncol. 2013;31(9):1134–9.

9 Radley DC, Finkelstein SN, Stafford RS. Off-label prescribing among office-based physicians. Arch Intern Med. 2006;166(9):1021–6.

10 Stephens T, Brynner R. Dark Remedy: The Impact of Thalidomide and Its Revival as a Vital Medicine. Basic Books; 2009. 244 p.

11 Anon. Israeli 'cures' leprosy. The Jewish Chronicle of Pittsburgh [Internet]. 1976;15(5):14. [cited 3 January 2014]. Available from: http://doi.library.cmu.edu/10.1184/pmc/CHR/CHR_1976_015_005_03111976. Accessed 8 August 2014.

12 Hampton T. Experts weigh in on promotion, prescription of off-label drugs. JAMA. 2007;297(7):683–5.

13 Smyth AR, Barbato A, Beydon N, et al. Respiratory medicines for children: Current evidence, unlicensed use and research priorities. Eur Respir J. 2010;35(2):247–65.

14 Lindell-Osuagwu L, Korhonen MJ, Saano S, et al. Off-label and unlicensed drug prescribing in three paediatric wards in Finland and review of the international literature. J Clin Pharm Ther. 2009;34(3):277–87.

15 Kimland E, Odlind V. Off-label drug use in pediatric patients. Clin Pharmacol Ther. 2012;91(5):796–801.

16 Horen B, Montastruc J-L, Lapeyre-mestre M. Adverse drug reactions and off-label drug use in paediatric outpatients. Br J Clin Pharmacol. 2002;54(6):665–70.

17 Ufer M, Kimland E, Bergman U. Adverse drug reactions and off-label prescribing for paediatric outpatients: A one-year survey of spontaneous reports in Sweden. Pharmacoepidemiol Drug Saf. 2004;13(3):147–52.

18 Turner S, Nunn AJ, Fielding K, et al. Adverse drug reactions to unlicensed and off-label drugs on paediatric wards: A prospective study. Acta Paediatr. 1999;88(9):965–8.

19 Choonara I, Conroy S. Unlicensed and off-label drug use in children: Implications for safety. Drug Saf Int J Med Toxicol Drug Exp. 2002;25(1):1–5.

20 Aagaard L, Hansen EH. Prescribing of medicines in the Danish paediatric population out-with the licensed age group: Characteristics of adverse drug reactions. Br J Clin Pharmacol. 2011;71(5):751–7.

21 Graziul C, Gibbons R, Alexander GC. Association between the commercial characteristics of psychotropic drugs and their off-label use. Med Care. 2012;50(11):940–7.

22 Banerjee S. The use of antipsychotic medication for people with dementia: Time for action. Dep Health. 2009. Available from: http://www.bmj.com/content/342/bmj.d3514. Accessed 8 August 2014.

23 Anon. Low-Dose Antipsychotics in People with Dementia [Internet]. NICE Evidence. March 2012 [cited 3 January 2014]. Available from: http://publications.nice.org.uk/low-dose-anti-psychotics-in-people-with-dementia-ktt7/evidence-context#close. Accessed 8 August 2014.

24 Harding R, Peel E. 'He was like a zombie': Off-label prescription of antipsychotic drugs in dementia. Med Law Rev. 2013;21(2):243–77.

25 Levinson D. Medicare atypical antipsychotic drug claims for elderly nursing home residents [Internet]. Department of Health and Human Services Office of Inspector General Report (OEI- 07-08-00150) May 2001 [cited 3 January 2014]. Available from: https://oig.hhs.gov/oei/reports/oei-07-08-00150.pdf. Accessed 8 August 2014.

26 Friedman RA. Wars on Drugs. The New York Times [Internet]. 6 April 2013 [cited 3 January 2014]. Available from: http://www.nytimes.com/2013/04/07/opinion/sunday/wars-on-drugs.html. Accessed 8 August 2014.

27 Krystal JH, Rosenheck RA, Cramer JA, et al. Adjunctive risperidone treatment for antide-pressant-resistant symptoms of chronic military service–related PTSD: A randomized trial. JAMA. 2011;306(5):493–502.

28 Lowe-Ponsford F, Baldwin D. Off-label prescribing by psychiatrists. Psychiatr Bull. 2000;24(11):415–7.

29 Weiss E, Hummer M, Koller D, et al. Off-label use of antipsychotic drugs. J Clin Psychopharmacol. 2000;20(6):695–8.

30 Hodgson R, Belgamwar R. Off-label prescribing by psychiatrists. Psychiatr Bull. 2006;30(2):55–7.

31 Devulapalli KK, Nasrallah HA. An analysis of the high psychotropic off-label use in psychi-atric disorders: The majority of psychiatric diagnoses have no approved drug. Asian J Psychiatry. 2009;2(1):29–36.

32 Anon. Off label prescribing turns out to be on label [Internet] 1 December 2008 [cited 3 January 2014]. Available from: http://thelastpsychiatrist.com/2008/12/off_label_pre-scribing.html. Accessed 8 August 2014.

33 Linder JA, Bates DW, Williams DH, et al. Acute infections in primary care: Accuracy of electronic diagnoses and electronic antibiotic prescribing. J Am Med Inform Assoc. 2006;13(1):61–6.

34 Chen DT, Wynia MK, Moloney RM, et al. U.S. physician knowledge of the FDA-approved indications and evidence base for commonly prescribed drugs: Results of a national survey. Pharmacoepidemiol Drug Saf. 2009;18(11):1094–100.

35 Baldwin DS, Kosky N. Off-label prescribing in psychiatric practice. Adv Psychiatr Treat. 2007;13(6):414–22.

36 Royal College of Psychiatrists. Use of licensed medicines for unlicensed applications in psychiatric practice. Psychiatr Bull. 2007;31(7):275.

37 Dennis M. US reviewing atypical antipsychotic use in children: Report [Internet] 12 August 2013 [cited 3 January 2014]. Available from: http://www.firstwordpharma.com/node/1130783. Accessed 8 August 2014.

38 Lagnado L. U.S. probes use of antipsychotic drugs on children. Wall Street J [Internet]. 12 August 2013 [cited 3 January 2014]. Available from: http://online.wsj.com/news/articles/SB10001424127887323477604578654130865747470. Accessed 8 August 2014.

39 Mercola JM. Why are antipsychotic drugs prescribed to children? [Internet]. Mercola.com 4 April 2012 [cited 3 January 2014]. Available from: http://articles.mercola.com/sites/articles/archive/2012/04/04/antipsychotic-drugs-on-pediatric-bipolar-disorder.aspx. Accessed 8 August 2014.

40 Casali PG. The off-label use of drugs in oncology: A position paper by the European Society for Medical Oncology (ESMO). Ann Oncol. 2007;18(12):1923–5.

41 Soares M. "Off-label" indications for oncology drug use and drug compendia: History and current status. J Oncol Pract. 2005;1(3):102–5.

42 Seaman A. Almost one-third of chemotherapy used 'off-label' [Internet]. ChrisBeatCancer. com [cited 2 August 2014]. Available from: http://www.reuters.com/article/2013/02/19/us-chemotherapy-idUSBRE91I18T20130219. Accessed 8 August 2014.

43 Eguale T, Buckeridge DL, Winslade NE, et al. Drug, patient, and physician characteristics associated with off-label prescribing in primary care. Arch Intern Med. 2012;172(10):781–8.

44 Gazarian M, Kelly M, McPhee JR, et al. Off-label use of medicines: Consensus recommendations for evaluating appropriateness. Med J Aust [Internet]. 2006 [cited 3 January 2014];185(10). Available from: https://www.mja.com.au/journal/2006/185/10/label-use-medicines-consensus-recommendations-evaluating-appropriateness. Accessed 8 August 2014.

45 Stafford RS. Off-label use of drugs and medical devices: A review of policy implications. Clin Pharmacol Ther. 2012;91(5):920–5.

46 Winslow R. Off-label use of clot drug is faulted. Wall Street J [Internet]. 19 April 2011 [cited 3 January 2014]. Available from: http://online.wsj.com/news/articles/SB10001424052748703916004576271202909607390. Accessed 8 August 2014.

47 Alten JA, Benner K, Green K, et al. Pediatric off-label use of recombinant factor VIIa. Pediatrics. 2009;123(3):1066–72.

48 Yank V, Tuohy CV, Logan AC, et al. Systematic review: Benefits and harms of in-hospital use of recombinant factor VIIa for off-label indications. Ann Intern Med. 2011;154(8):529–40.

49 Avorn J, Kesselheim A. A hemorrhage of off-label use. Ann Intern Med. 2011;154(8):566–7.

50 Fauber J. Complications rise along with off-label use of BMP-2 [Internet] 28 August 2010 [cited 3 January 2014]. Available from: http://www.jsonline.com/news/health/101732923.html. Accessed 8 August 2014.

51 United States Senate Finance Committee. Staff report on Medtronic's influence on INFUSE clinical studies. Int J Occup Environ Health. 2013;19(2):67–76.

52 Krumholz HM, Ross JS, Gross CP, et al. A historic moment for open science: The Yale University open data access project and Medtronic. Ann Intern Med. 2013;158(12):910–1.

53 Epstein NE. Complications due to the use of BMP/INFUSE in spine surgery: The evidence continues to mount. Surg Neurol Int. 2013;4(Supplement 5):S343–52.

54 Fauber J, Reporter, Today MJS. Big bucks, no benefits with many drugs [Internet] 9 March 2013 [cited 3 January 2014]. Available from: http://www.medpagetoday.com/Cardiology/Dyslipidemia/37772. Accessed 8 August 2014.

55 Peñaloza RA, Sarkar U, Claman DM, et al. Trends in on-label and off-label modafinil use in a nationally representative sample. JAMA Intern Med. 2013;173(8):704–6.

56 Peterson RE, Imperato-McGinley J, Gautier T, et al. Male pseudohermaphroditism due to steroid 5α-reductase deficiency. Am J Med. 1977;62(2):170–91.

57 Golub LM, Ciancio S, Ramamamurthy NS, et al. Low-dose doxycycline therapy: Effect on gingival and crevicular fluid collagenase activity in humans. J Periodontal Res. 1990;25(6):321–30.

58 Buzdar AU, Marcus C, Holmes F, et al. Phase II evaluation of Ly156758 in metastatic breast cancer. Oncology. 1988;45(5):344–5.

59 Eastell R. Treatment of postmenopausal osteoporosis. N Engl J Med. 1998;338(11):736–46.

60 Heinrich M, Lee Teoh H. Galanthamine from snowdrop—the development of a modern drug against Alzheimer's disease from local Caucasian knowledge. J Ethnopharmacol. 2004;92(2–3):147–62.

61 Heusler K, Pletscher A. The controversial early history of cyclosporin. Swiss Med Wkly. 2001;131(21–22):299–302.

62 Barber J. EU committee says Biogen Idec's MS drug Tecfidera qualifies as new active substance [Internet] 22 November 2013 [cited 3 January 2014]. Available from: http://www.firstwordpharma.com/node/1165633. Accessed 8 August 2014.

63 Dharmshaktu P, Tayal V, Kalra BS. Efficacy of antidepressants as analgesics: A review. J Clin Pharmacol. 2012;52(1):6–17.

64 Olmsted CL, Kockler DR. Topiramate for alcohol dependence. Ann Pharmacother. 2008;42(10):1475–80.

65 Rothwell PM, Fowkes FGR, Belch JF, *et al.* Effect of daily aspirin on long-term risk of death due to cancer: Analysis of individual patient data from randomised trials. Lancet. 2011;377(9759):31–41.

66 Cavalla D, Singal C. Retrospective clinical analysis for drug rescue: For new indications or stratified patient groups. Drug Discov Today. 2012;17(3–4):104–9.

67 Leone A, Di Gennaro E, Bruzzese F, *et al.* New perspective for an old antidiabetic drug: Metformin as anticancer agent. Cancer Treat Res. 2014;159:355–76.

68 Anon. Harris Interactive. News room – U.S. adults ambivalent about the risks and benefits of off-label prescription drug use [Internet] 7 December 2006 [cited 3 January 2014]. Available from: http://www.harrisinteractive.com/news/allnewsbydate.asp?NewsID=1126. Accessed 8 August 2014.

69 Anon. Harris Interactive. News room – Many people think that drugs should only be prescribed per FDA-approved use, not for off-label use [Internet] 9 June 2004 [cited 3 January 2014]. Available from: http://www.harrisinteractive.com/news/allnewsbydate.asp?NewsID=808. Accessed 8 August 2014.

70 Mukattash TL, Millership JS, Collier PS, *et al.* Public awareness and views on unlicensed use of medicines in children. Br J Clin Pharmacol. 2008;66(6):838–45.

71 Mukattash T, Trew K, Hawwa AF, *et al.* Children's views on unlicensed/off-label paediatric prescribing and paediatric clinical trials. Eur J Clin Pharmacol. 2012;68(2):141–8.

72 Goulding MR. Trends in prescribed medicine use and spending by older Americans 1992–2001 [Internet]. National Center for Health Statistics. February 2005 [cited 3 January 2014]. Available from: http://www.cdc.gov/nchs/data/ahcd/agingtrends/05medicine.pdf. Accessed 8 August 2014.

73 Bell MDD. The UK human tissue act and consent: Surrendering a fundamental principle to transplantation needs? J Med Ethics. 2006;32(5):283–6.

74 Anon. Alder Hey organs scandal [Internet]. Wikipedia, the free encyclopedia. 3 November 2013 [cited 22 March 2014]. Available from: http://en.wikipedia.org/w/index.php?title=Alder_Hey_organs_scandal&oldid=579945434. Accessed 8 August 2014.

75 Parsons S, Winterbottom A, Cross P, Redding D. The quality of patient engagement and involvement in primary care [Internet] 2011 [cited 3 January 2014]. Available from: http://www.kingsfund.org.uk/projects/gp-inquiry/patient-engagement-involvement. Accessed 8 August 2014.

76 Hartzband P, Groopman J. Untangling the web – Patients, doctors, and the internet. N Engl J Med. 2010;362(12):1063–6.

77 Frost J, Okun S, Vaughan T, *et al.* Patient-reported outcomes as a source of evidence in off-label prescribing: Analysis of data from PatientsLikeMe. J Med Internet Res. 2011;13(1):e6.

78 Rich BA. Off-label prescribing: In search of a reasonable patient-centered approach. J Pain Palliat Care Pharmacother. 2012;26(2):131–3.

79 General Medical Council. Good practice in prescribing and managing medicines and devices [Internet]. General Medical Council. 2013 [cited 3 January 2014]. Available from: http://www.gmc-uk.org/Prescribing_Guidance__2013__50955425.pdf. Accessed 8 August 2014.

80 Colyer S. Off-label guide welcomed. Med J Aust [Internet] 2014 [cited 4 August 2014]. Available from: https://www.mja.com.au/insight/2014/21/label-guide-welcomed. Accessed 8 August 2014.

81 Ray P. Report of the Council on Ethical and Judicial Affairs CEJA report 2-a-06 [Internet]. American Medical Association. 2006 [cited 3 January 2014]. Available from: http://www.ama-assn.org/resources/doc/ethics/x-pub/ceja_2a06.pdf. Accessed 8 August 2014.

82 Anon. AMA's Code of Medical Ethics [Internet]. American Medical Association [cited 3 January 2014]. Available from: http://www.ama-assn.org/ama/pub/physician-resources/medical-ethics/code-medical-ethics.page. Accessed 8 August 2014.

83 Elwyn G, Frosch D, Thomson R, *et al.* Shared decision making: A model for clinical practice. J Gen Intern Med. 2012;27(10):1361–7.

84 Wilkes M, Johns M. Informed consent and shared decision-making: A requirement to disclose to patients off-label prescriptions. PLoS Med [Internet]. November 2008 [cited 1 September 2013];5(11):e223. Available from: http://www.ncbi.nlm.nih.gov/pmc/articles/PMC2581625/. Accessed 8 August 2014.

85 Informed Medical Decisions Foundation. State legislative and regulatory approaches to shared decision making [Internet]. Informed Medical Decisions Foundation. March 2012 [cited 3 January 2014]. Available from: http://www.nashp.org/sites/default/files/shared.decision.making.companion.document.pdf. Accessed 8 August 2014.

86 Chisholm A. Exploring UK attitudes towards unlicensed medicines use: A questionnaire-based study of members of the general public and physicians. Int J Gen Med. 2012;5:27–40.

87 Ekins-Daukes S, Helms PJ, Taylor MW, et al. Off-label prescribing to children: Attitudes and experience of general practitioners. Br J Clin Pharmacol. 2005;60(2):145–9.

88 Stewart D, Rouf A, Snaith A, et al. Attitudes and experiences of community pharmacists towards paediatric off-label prescribing: A prospective survey. Br J Clin Pharmacol. 2007;64(1):90–5.

89 Mukattash TL, Wazaify M, Khuri-Boulos N, et al. Perceptions and attitudes of Jordanian paediatricians towards off-label paediatric prescribing. Int J Clin Pharm. 2011;33(6):964–73.

90 Ditsch N, Kumper C, Summerer-Moustaki M, et al. Off-label use in Germany – A current appraisal of gynaecologic university departments. Eur J Med Res. 2011;16(1):7–12.

91 Beck/Herrmann. Drug and device law: Informed consent and FDA regulatory status – Oil and water still don't mix [Internet] 21 June 2007 [cited 3 January 2014]. Available from: http://druganddevicelaw.blogspot.co.uk/2007/06/informed-consent-and-fda-regulatory.html. Accessed 8 August 2014.

92 FDA panel: Avastin ineffective for breast cancer [Internet]. Wellness.com [cited 3 January 2014]. Available from: http://www.wellness.com/news/9152/fda-panel-avastin-ineffective-for-breast-cancer/health-and-wellness-news. Accessed 8 August 2014.

93 Hacking I. Lost in the forest. London Rev Books. 2013;35(15):7–8.

94 Open to interpretation. Nat Biotechnol. 2013;31(8):661–661.

95 The good practice guidelines for GP electronic patient records [Internet]. Department of Health (DH)/Royal College of General Practitioners (RCGP)/British Medical Association (BMA); 2011 [cited 3 January 2014]. Available from: https://www.gov.uk/government/uploads/system/uploads/attachment_data/file/215680/dh_125350.pdf. Accessed 8 August 2014.

96 Lichtenberg FR. The impact of new drug launches on longevity: Evidence from longitudinal disease-level data from 52 countries, 1982–2001 [Internet]. National Bureau of Economic Research; June 2003. Report No.: 9754 [cited 23 March 2014]. Available from: http://www.nber.org/papers/w9754. Accessed 8 August 2014.

97 Paul SM, Mytelka DS, Dunwiddie CT, et al. How to improve R&D productivity: The pharmaceutical industry's grand challenge. Nat Rev Drug Discov. 2010;9(3):203–14.

98 Herper M. The truly staggering cost of inventing new drugs [Internet]. Forbes [cited 19 March 2014]. Available from: http://www.forbes.com/sites/matthewherper/2012/02/10/the-truly-staggering-cost-of-inventing-new-drugs/. Accessed 8 August 2014.

99 Hay M, Thomas DW, Craighead JL, et al. Clinical development success rates for investigational drugs. Nat Biotechnol. 2014;32(1):40–51.

100 Thayer A. Drug repurposing. Chemical & Engineering News [Internet]. 1 October 2012 [cited 20 March 2014]; 90(40). Available from: http://cen.acs.org/articles/90/i40/Drug-Repurposing.html?h=-1031248274. Accessed 8 August 2014.

101 Anon. News in brief: Pfizer's Dimebon deal. Nat Rev Drug Discov. 2008;7(10):792–3.

102 Cha M, Rifai B, Sarraf P. Pharmaceutical forecasting: Throwing darts? Nat Rev Drug Discov. 2013;12(10):737–8.

103 Butler J. Value line – Dendreon Corp [Internet] 11 December 2010 [cited 3 January 2014]. Available from: http://www.valueline.com/Stocks/Highlight.aspx?id=9982. Accessed 8 August 2014.

104 Bombardier C, Laine L, Reicin A, et al. Comparison of upper gastrointestinal toxicity of rofecoxib and naproxen in patients with rheumatoid arthritis. N Engl J Med. 2000;343(21):1520–8.

105 Mukherjee D, Nissen SE, Topol EJ. Risk of cardiovascular events associated with selective cox-2 inhibitors. JAMA. 2001;286(8):954–9.

106 Whoriskey P. Amgen and its erythropoietin drugs [Internet]. The Washington Post. 19 July 2012 [cited 20 February 2014]. Available from: http://www.washingtonpost.com/wp-srv/business/amgen-anemia-drugs/index.html. Accessed 8 August 2014.

107 Kesselheim AS, Myers JA, Solomon DH, *et al*. The prevalence and cost of unapproved uses of top-selling orphan drugs. PloS One. 2012;7(2):e31894.

108 Gagnon M-A, Lexchin J. The cost of pushing pills: A new estimate of pharmaceutical promotion expenditures in the United States. PLoS Med. 2008;5(1):e1.

109 Steinman MA, Bero LA, Chren M-M, *et al*. Narrative review: The promotion of Gabapentin: An analysis of internal industry documents. Ann Intern Med. 2006;145(4):284–93.

110 Steinman M, Landefeld CS, Chren M. United States ex rel. Franklin v. Parke-Davis, Expert Consultant's Report [estimates of gabapentin uses, by diagnosis.] [cited 3 January 2014] . Available from http://dida.library.ucsf.edu/. Accessed 8 August 2014.

111 Boodman S. Off-label use of risky antipsychotic drugs raises concerns – Schizophrenia center [Internet]. EverydayHealth.com [cited 7 January 2014]. Available from: http://www.everydayhealth.com/schizophrenia/0313/off-label-use-of-risky-antipsychotic-drugs-raises-concerns.aspx. Accessed 8 August 2014.

112 Frosch DL, Grande D, Tarn DM, *et al*. A decade of controversy: Balancing policy with evidence in the regulation of prescription drug advertising. Am J Public Health. 2010;100(1):24–32.

113 DiMasi JA, Paquette C. The economics of follow-on drug research and development: Trends in entry rates and the timing of development. PharmacoEconomics. 2004;22(2 Supplement 2):1–14.

114 Cavalla D. APT drug R&D: The right active ingredient in the right presentation for the right therapeutic use. Nat Rev Drug Discov. 2009;8(11):849–53.

115 Bellis JR, Kirkham JJ, Thiesen S, *et al*. Adverse drug reactions and off-label and unlicensed medicines in children: A nested case? Control study of inpatients in a pediatric hospital. BMC Med. 2013;11(1):238.

116 Santos DB, Clavenna A, Bonati M, *et al*. Off-label and unlicensed drug utilization in hospitalized children in Fortaleza, Brazil. Eur J Clin Pharmacol. 2008;64(11):1111–8.

117 Clarkson A, Conroy S, Burroughs K, *et al*. Surveillance for adverse drug reactions in children: A paediatric regional monitoring centre. Paediatr Perinat Drug Ther. 2004;6(1):20–3.

118 Evidence of harm from off-label or unlicensed medicines in children EMEA [Internet]. European Medicines Agency; 2004 [cited 12 January 2014]. Available from: http://www.ema.europa.eu/docs/en_GB/document_library/Other/2009/10/WC500004021.pdf. Accessed 8 August 2014.

119 Jonville-Béra AP, Béra F, Autret-Leca E. Are incorrectly used drugs more frequently involved in adverse drug reactions? A prospective study. Eur J Clin Pharmacol. 2005;61(3):231–6.

120 Stafford RS. Regulating off-label drug use – Rethinking the role of the FDA. N Engl J Med. 2008;358(14):1427–9.

121 Jonville-Béra AP, Saissi H, Bensouda-Grimaldi L, *et al*. Avoidability of adverse drug reactions spontaneously reported to a French regional drug monitoring centre. Drug Saf Int J Med Toxicol Drug Exp. 2009;32(5):429–40.

122 Ohlow M-A, von Korn H, Gunkel O, *et al*. Incidence of adverse cardiac events 5 years after polymer-free sirolimus eluting stent implantation: Results from the prospective Bad Berka Yukon Choice™ registry. Catheter Cardiovasc Interv. 2013. DOI:10.1002/ccd.25272.

123 Bridges J, Maisel WH. Malfunctions and adverse events associated with off-label use of biliary stents in the peripheral vasculature. Am J Ther. 2008;15(1):12–8.

124 Mason J, Pirmohamed M, Nunn T. Off-label and unlicensed medicine use and adverse drug reactions in children: A narrative review of the literature. Eur J Clin Pharmacol. 2012;68(1):21–8.

125 Off-base: The exclusion of off-label prescriptions from Medicare Part D coverage [Internet]. Medicare Rights Center; 2007 [cited 3 January 2014]. Available from: http://www.policy-archive.org/handle/10207/8863. Accessed 8 August 2014.

126 Paulozzi LJ, Jones CM, Mack KA, Rudd RA. Vital signs: Overdoses of prescription opioid pain relievers – United States, 1999–2008 [Internet] 4 November 2011 [cited 6 January 2014]. Available from: http://www.cdc.gov/mmwr/preview/mmwrhtml/mm6043a4.htm. Accessed 8 August 2014.

127 Voice FDA. Why FDA supports a flexible approach to drug development. FDA voice [Internet] [cited 7 February 2014]. Available from: http://blogs.fda.gov/fdavoice/index.php/2014/02/why-fda-supports-a-flexible-approach-to-drug-development/. Accessed 8 August 2014.

128 Hamer M, Batty GD, David Batty G, et al. Antidepressant medication use and future risk of cardiovascular disease: The Scottish Health Survey. Eur Heart J. 2011;32(4):437–42.

129 Revised February 27, 2006, by the American Society of Clinical Oncology. Reimbursement for cancer treatment: Coverage of off-label drug indications. J Clin Oncol. 2006;24(19): 3206–8.

130 Gavin PJ, Thomson Jr. RB. Review of rapid diagnostic tests for influenza. Clin Appl Immunol Rev. 2004;4(3):151–72.

131 Arroll B, Kenealy T. Antibiotics for the common cold and acute purulent rhinitis. Cochrane Database Syst Rev [Internet]. John Wiley & Sons, Ltd; 1996 [cited 3 January 2014]. Available from: http://onlinelibrary.wiley.com/doi/10.1002/14651858.CD000247/abstract. Accessed 8 August 2014.

132 Smucny J, Fahey T, Becker L, et al. Antibiotics for acute bronchitis. Cochrane Database Syst Rev. 2004;(4):CD000245.

133 Ball J, Hickman L, Campbell D. Antibiotic prescription by GP practice. The Guardian [Internet] [cited 18 February 2014]. Available from: http://www.theguardian.com/society/datablog/2013/jun/11/antibiotic-prescription-by-gp-practice-mapped. Accessed 8 August 2014.

134 Fendrick A, Monto AS, Nightengale B, et al. The economic burden of non–influenza-related viral respiratory tract infection in the united states. Arch Intern Med. 2003;163(4):487–94.

135 Etminan M, Forooghian F, Brophy JM, et al. Oral fluoroquinolones and the risk of retinal detachment. JAMA. 2012;307(13):1414–9.

136 Bird ST, Etminan M, Brophy JM, et al. Risk of acute kidney injury associated with the use of fluoroquinolones. Can Med Assoc J. 2013;185(10):E475–82.

137 Brody JE. Popular antibiotics may carry serious side effects [Internet]. Well. 10 September 2012 [cited 3 January 2014]. Available from: http://well.blogs.nytimes.com/2012/09/10/popular-antibiotics-may-carry-serious-side-effects/. Accessed 8 August 2014.

138 Pépin J, Saheb N, Coulombe M-A, et al. Emergence of fluoroquinolones as the predominant risk factor for Clostridium difficile-associated diarrhea: A cohort study during an epidemic in Quebec. Clin Infect Dis Off Publ Infect Dis Soc Am. 2005;41(9):1254–60.

139 Flowers CM, Racoosin JA, Kortepeter C. Seizure activity and off-label use of tiagabine. N Engl J Med. 2006;354(7):773–4.

140 Fugh-Berman A, Melnick D. Off-label promotion, on-target sales. PLoS Med. 2008;5(10):e210.

141 Lucak SL. Optimizing outcomes with alosetron hydrochloride in severe diarrhea-predominant irritable bowel syndrome. Ther Adv Gastroenterol. 2010;3(3):165–72.

142 Connolly HM, Crary JL, McGoon MD, et al. Valvular heart disease associated with fenfluramine–phentermine. N Engl J Med. 1997;337(9):581–8.

143 MacGregor EA, Brandes JL, Silberstein S, et al. Safety and tolerability of short-term preventive frovatriptan: A combined analysis. Headache. 2009;49(9):1298–314.

144 Kaplan S, Staffa JA, Dal Pan GJ. Duration of therapy with metoclopramide: A prescription claims data study. Pharmacoepidemiol Drug Saf. 2007;16(8):878–81.

145 Mesgarpour B, Heidinger BH, Schwameis M, et al. Safety of off-label erythropoiesis stimulating agents in critically ill patients: A meta-analysis. Intensive Care Med. 2013;39(11): 1896–908.

146 Astrup A, Rössner S, Van Gaal L, et al. Effects of liraglutide in the treatment of obesity: A randomised, double-blind, placebo-controlled study. Lancet. 2009;374(9701):1606–16.

147 LaPointe N, Chen AY, Alexander KP, et al. Enoxaparin dosing and associated risk of in-hospital bleeding and death in patients with non–ST-segment elevation acute coronary syndromes. Arch Intern Med. 2007;167(14):1539–44.

148 Stiers JL, Ward RM. Newborns, one of the last therapeutic orphans to be adopted. JAMA Pediatr. 2014;168(2):106–8.

149 Schirm E, Tobi H, Berg LTW de J den. Risk factors for unlicensed and off-label drug use in children outside the hospital. Pediatrics. 2003;111(2):291–5.

150 Abman SH, Kinsella JP, Rosenzweig EB, et al. Implications of the U.S. Food and Drug Administration warning against the use of sildenafil for the treatment of pediatric pulmonary hypertension. Am J Respir Crit Care Med. 2013;187(6):572–5.

151 Zito JM, Derivan AT, Kratochvil CJ, et al. Off-label psychopharmacologic prescribing for children: History supports close clinical monitoring. Child Adolesc Psychiatry Ment Health. 2008;2(1):24.

152 Busen NH, Britt RB, Rianon N. Bone mineral density in a cohort of adolescent women using depot medroxyprogesterone acetate for one to two years. J Adolesc Health. 2003;32(4):257–9.

153 Carpenter DJ, Fong R, Kraus JE, et al. Meta-analysis of efficacy and treatment-emergent suicidality in adults by psychiatric indication and age subgroup following initiation of paroxetine therapy: A complete set of randomized placebo-controlled trials. J Clin Psychiatry. 2011;72(11):1503–14.

154 Simon GE, Gregory, Savarino J. Suicide attempts among patients starting depression treatment with medications or psychotherapy. Am J Psychiatry. 2007;164(7):1029–34.

155 Lu CY, Zhang F, Lakoma MD, et al. Changes in antidepressant use by young people and suicidal behavior after FDA warnings and media coverage: quasi-experimental study. BMJ. 2014;348:g3596–g3596.

156 Gibbons RD, Brown CH, Hur K, et al. Suicidal thoughts and behavior with antidepressant treatment. Arch Gen Psychiatry. 2012;69(6):580–7.

157 Bücheler R, Schwab M, Mörike K, et al. Off label prescribing to children in primary care in Germany: Retrospective cohort study. BMJ. 2002;324(7349):1311–2.

158 Weeks AD, Fiala C, Safar P. Misoprostol and the debate over off-label drug use. BJOG Int J Obstet Gynaecol. 2005;112(3):269–72.

159 Maher A, Maglione M, Bagley S, et al. Efficacy and comparative effectiveness of atypical antipsychotic medications for off-label uses in adults: A systematic review and meta-analysis. JAMA. 2011;306(12):1359–69.

160 Lazarou J, Pomeranz BH, Corey PN. Incidence of adverse drug reactions in hospitalized patients: A meta-analysis of prospective studies. JAMA. 1998;279(15):1200–5.

161 Hakkarainen KM, Hedna K, Petzold M, et al. Percentage of patients with preventable adverse drug reactions and preventability of adverse drug reactions–A meta-analysis. PLoS ONE. 2012;7(3):e33236.

162 Katzberg HD, Khan AH, So YT. Assessment: Symptomatic treatment for muscle cramps (an evidence-based review) report of the Therapeutics and Technology Assessment Subcommittee of the American Academy of neurology. Neurology 2010;74(8):691–6.

163 Lautenbach E, Larosa LA, Kasbekar N, et al. Fluoroquinolone utilization in the emergency departments of academic medical centers: Prevalence of, and risk factors for, inappropriate use. Arch Intern Med. 2003;163(5):601–5.

164 Cheng K, Masters S, Stephenson T, et al. Identification of suspected fatal adverse drug reactions by paediatricians: A UK surveillance study. Arch Dis Child. 2008;93(7): 609–11.

165 Levine I. Off-label drugs, FDA, prescription medications, drugmakers – AARP bulletin [Internet]. AARP. April 2008 [cited 17 February 2014]. Available from: http://www.aarp. org/health/drugs-supplements/info-04-2009/off-label_drugs__what.html. Accessed 8 August 2014.

166 Von Haehling S, Anker SD. beta-blockers in heart failure – Much more than heart rate reduction. J Card Fail. 2002;8(6):379–80.

167 Cavalla D. Treatment of cachexia. Patent EP2094254 B1, 2011.

168 Wittich CM, Burkle CM, Lanier WL. Ten common questions (and their answers) about off-label drug use. Mayo Clin Proc. 2012;87(10):982–90.

169 Montague DK, Jarow J, Broderick GA, et al. AUA guideline on the pharmacologic management of premature ejaculation. J Urol. 2004;172(1):290–4.

170 Seely DMR, Wu P, Mills EJ. EDTA chelation therapy for cardiovascular disease: A systematic review. BMC Cardiovasc Disord. 2005;5:32.

171 Walton S, Schumock G, Lee K-V. Developing evidence-based research priorities for off-label drug use [Internet]. Effective Health Care Research Report No. 12. (Prepared by the University of Illinois at Chicago DEcIDE Center Under Contract No. HHSA290200500038I T03.) Rockville, MD: Agency for Healthcare Research and Quality. May 2009 [cited 12 January 2014]. Available from: http://effectivehealthcare.ahrq.gov/reports/final.cfm. Accessed 8 August 2014.

172 Saad M, Cassagnol M, Ahmed E. The impact of FDA's warning on the use of antipsychotics in clinical practice: A survey. Consult Pharm J Am Soc Consult Pharm. 2010; 25(11):739–44.

173 Gleason PP, Walters C, Heaton AH, et al. Telithromycin: The perils of hasty adoption and persistence of off-label prescribing. J Manag Care Pharm. 2007;13(5):420–5.

174 Anon. Information NC for B, Pike USNL of M 8600 R, MD B, USA 20894. Polycystic ovary syndrome: Does the antidiabetic drug metformin increase fertility? [Internet]. 19 July 2012 [cited 18 February 2014]. Available from: https://www.ncbi.nlm.nih.gov/pubmed-health/PMH0048142/. Accessed 8 August 2014.

175 Barnes PJ. Inhaled corticosteroids in COPD: A controversy. Respir Int Rev Thorac Dis. 2010;80(2):89–95.

176 United States General Accounting Office: Off- label drugs: Reimbursement policies constrain physicians in their choice of cancer therapies [Internet] [cited 3 January 2014]. GAO/PEMD-91-14; 1991. Available from: http://www.gao.gov/assets/160/151121.pdf. Accessed 8 August 2014.

177 Kocs D, Fendrick AM. Effect of off-label use of oncology drugs on pharmaceutical costs: The rituximab experience. Am J Manag Care. 2003;9(5):393–400; quiz 401–2.

178 Krzyzanowska MK. Off-label use of cancer drugs: A benchmark is established. J Clin Oncol. 2013;31(9):1125–7.

179 Schwartz LM, Woloshin S. Low 'T' as in 'template': How to sell disease. JAMA Intern Med. 2013;173(15):1460–2.

180 Martinez FD. Children, asthma, and proton pump inhibitors: Costs and perils of therapeutic creep. JAMA. 2012;307(4):406–7.

181 Largent EA, Miller FG, Pearson SD. Going off-label without venturing off-course: Evidence and ethical off-label prescribing. Arch Intern Med. 2009;169(19):1745–7.

182 Keatings VM, Collins PD, Scott DM, et al. Differences in interleukin-8 and tumor necrosis factor-alpha in induced sputum from patients with chronic obstructive pulmonary disease or asthma. Am J Respir Crit Care Med. 1996;153(2):530–4.

183 Vernooy JH, Küçükaycan M, Jacobs JA, et al. Local and systemic inflammation in patients with chronic obstructive pulmonary disease: Soluble tumor necrosis factor receptors are increased in sputum. Am J Respir Crit Care Med. 2002;166(9):1218–24.

184 Churg A, Wang RD, Tai H, et al. Tumor necrosis factor-alpha drives 70% of cigarette smoke-induced emphysema in the mouse. Am J Respir Crit Care Med. 2004;170(5):492–8.

185 Zhang X-Y, Zhang C, Sun Q-Y, et al. Infliximab protects against pulmonary emphysema in smoking rats. Chin Med J (Engl). 2011;124(16):2502–6.

186 Rennard SI, Fogarty C, Kelsen S, et al. The safety and efficacy of infliximab in moderate to severe chronic obstructive pulmonary disease. Am J Respir Crit Care Med. 2007;175(9):926–34.

187 Jit M, Henderson B, Stevens M, et al. TNF-alpha neutralization in cytokine-driven diseases: A mathematical model to account for therapeutic success in rheumatoid arthritis but therapeutic failure in systemic inflammatory response syndrome. Rheumatol Oxf Engl. 2005;44(3):323–31.

188 TNF neutralization in MS: Results of a randomized, placebo-controlled multicenter study. The Lenercept Multiple Sclerosis Study Group and The University of British Columbia MS/MRI Analysis Group. Neurology. 1999;53(3):457–65.

189 Sangle SR, Hughes GRV, D'Cruz DP. Infliximab in patients with systemic vasculitis that is difficult to treat: Poor outcome and significant adverse effects. Ann Rheum Dis. 2007;66(4):564–5.

190 Mariette X, Ravaud P, Steinfeld S, *et al.* Inefficacy of infliximab in primary Sjögren's syndrome: Results of the randomized, controlled Trial of Remicade in Primary Sjögren's Syndrome (TRIPSS). Arthritis Rheum. 2004;50(4):1270–6.

191 Wegener's Granulomatosis Etanercept Trial (WGET) Research Group. Etanercept plus standard therapy for Wegener's granulomatosis. N Engl J Med. 2005;352(4):351–61.

192 Cohen S, Shoup A, Weisman MH, *et al.* Etanercept treatment for autoimmune inner ear disease: Results of a pilot placebo-controlled study. Otol Neurotol Off Publ Am Otol Soc Am Neurotol Soc Eur Acad Otol Neurotol. 2005;26(5):903–7.

193 Hoffman GS, Cid MC, Rendt-Zagar KE, *et al.* Infliximab for maintenance of glucocorticosteroid-induced remission of giant cell arteritis: A randomized trial. Ann Intern Med. 2007;146(9):621–30.

194 Chung ES, Packer M, Lo KH, *et al.* Anti-TNF therapy against congestive heart failure investigators. Randomized, double-blind, placebo-controlled, pilot trial of infliximab, a chimeric monoclonal antibody to tumor necrosis factor-alpha, in patients with moderate-to-severe heart failure: Results of the anti-TNF Therapy Against Congestive Heart Failure (ATTACH) trial. Circulation. 2003;107(25):3133–40.

195 Carroll J. Managing the immunomodulators. Biotechnol Healthc. 2010;7(2):12–6.

196 Utz JP, Limper AH, Kalra S, *et al.* Etanercept for the treatment of stage II and III progressive pulmonary sarcoidosis. Chest. 2003;124(1):177–85.

197 Ohshiro K, Sakima A, Nakada S, *et al.* Beneficial effect of switching from a combination of angiotensin II receptor blockers other than losartan and thiazides to a fixed dose of losartan/hydrochlorothiazide on uric acid metabolism in hypertensive patients. Clin Exp Hypertens N Y N 1993. 2011;33(8):565–70.

198 Solomon DH, Avorn J, Stürmer T, *et al.* Cardiovascular outcomes in new users of coxibs and nonsteroidal antiinflammatory drugs: High-risk subgroups and time course of risk. Arthritis Rheum. 2006;54(5):1378–89.

199 Coxib and traditional NSAID Trialists' (CNT) Collaboration. Vascular and upper gastrointestinal effects of non-steroidal anti-inflammatory drugs: Meta-analyses of individual participant data from randomised trials. Lancet. 2013;382(9894):769–79.

200 Frech EJ, Go MF. Treatment and chemoprevention of NSAID-associated gastrointestinal complications. Ther Clin Risk Manag. 2009;5:65–73.

201 Prasad V, Vandross A, Toomey C, *et al.* A decade of reversal: An analysis of 146 contradicted medical practices. Mayo Clin Proc. 2013;88(8):790–8.

202 Eder JP, Antman K, Peters W, *et al.* High-dose combination alkylating agent chemotherapy with autologous bone marrow support for metastatic breast cancer. J Clin Oncol Off J Am Soc Clin Oncol. 1986;4(11):1592–7.

203 Jacobson PD, Rettig RA, Aubry WM. Litigating the science of breast cancer treatment. J Health Polit Policy Law. 2007;32(5):785–818.

204 Mullard A. Mediator scandal rocks French medical community. Lancet. 2011;377(9769):890–2.

205 Frachon I, Etienne Y, Jobic Y, *et al.* Benfluorex and unexplained valvular heart disease: A case-control study. PLoS ONE. 2010;5(4):e10128.

206 Jackson PSHD, Jansen PAF, Mangoni AA. Off-label prescribing in older patients. Drugs Aging. 2012;29(6):427–34.

207 MacDonald R. GlaxoSmithKline found guilty while complicit physicians remain unscathed [Internet]. Speaking of Medicine. 2012 [cited 23 January 2014]. Available from: http://blogs.plos.org/speakingofmedicine/2012/07/09/glaxosmithkline-found-guilty-while-complicit-physicians-remain-unscathed/. Accessed 8 August 2014.

208 Weber T. ProPublica CO, 3 J, 2012, p.m. 1:55. Drug companies reduce payments to doctors as scrutiny mounts [Internet]. ProPublica [cited 23 January 2014]. Available from: http://www.propublica.org/article/drug-companies-reduce-payments-to-doctors-as-scrutiny-mounts. Accessed 8 August 2014.

209 Staton T. More scrutiny, less pay? U.K. docs collected less cash from pharma in 2013 [Internet]. FiercePharma [cited 3 April 2014]. Available from: http://www.fiercepharma.com/story/more-scrutiny-less-pay-uk-docs-collected-less-cash-pharma-2013/2014-04-03. Accessed 8 August 2014.

210 Vedula SS, Li T, Dickersin K. Differences in reporting of analyses in internal company documents versus published trial reports: Comparisons in industry-sponsored trials in off-label uses of gabapentin. PLoS Med. 2013;10(1):e1001378.

211 Fugh-Berman A. How basic scientists help the pharmaceutical industry market drugs. PLoS Biol [Internet]. November 2013 [cited 15 January 2014];11(11):e1001716. Available from: http://www.ncbi.nlm.nih.gov/pmc/articles/PMC3833865/. Accessed 8 August 2014.

212 Molyneux CG, Bogaert P. The need for informed consent in off-label use in the EU. SCRIP Regul Aff. 2010;November:13–6.

213 Risk management when using drugs or medical devices off-label [Internet]. Canadian Medical Protective Association. 2012 [cited 3 January 2014]. Available from: https://oplfrpd5.cmpa-acpm.ca/en/duties-and-responsibilities/-/asset_publisher/bFaUiyQG069N/content/risk-management-when-using-drugs-or-medical-devices-off-label. Accessed 8 August 2014.

214 Riley JB Jr, Basilius PA. Physicians' liability for off-label prescriptions. Nephrol News Issues. 2007;21(7):43–4, 46–7.

215 Beck JM, Azari ED. FDA, off-label use, and informed consent: Debunking myths and misconceptions. Food Drug Law J. 1998;53(1):71–104.

216 Hill P. Off licence and off label prescribing in children: Litigation fears for physicians. Arch Dis Child. 2005;90(Supplement 1):ii7–8.

217 Koyuncu A. Arzneimittelversorgung im off-label-use – der rechtliche Rahmen. DMW – Dtsch Med Wochenschr. 2012;137(30):1519–23.

218 Staton T. Express scripts assembling anti-Sovaldi coalition to shut out Gilead hep C drug [Internet]. FiercePharma. 2014 [cited 9 April 2014]. Available from: http://www.fiercepharma.com/story/express-scripts-assembling-anti-sovaldi-coalition-shut-out-gilead-hep-c-dru/2014-04-08. Accessed 8 August 2014.

219 Miller MD. Sovaldi and curing hep C – myths and other facts|health policy and communications [Internet] 2014 [cited 5 August 2014]. Available from: http://www.healthpolcom.com/blog/2014/06/16/sovaldi-and-curing-hep-c-myths-and-other-facts/. Accessed 8 August 2014.

220 Levêque D. Off-label use of anticancer drugs. Lancet Oncol. 2008;9(11):1102–7.

221 Abernethy AP, Raman G, Balk EM, *et al.* Systematic review: Reliability of compendia methods for off-label oncology indications. Ann Intern Med. 2009;150(5):336–43.

222 Mullins CD, Montgomery R, Tunis S. Uncertainty in assessing value of oncology treatments. Oncologist. 2010;15(Supplement 1):58–64.

223 EFPIA. Promotion of off-label use of medicines by European healthcare bodies in indications where authorised medicines are available [Internet] May 2014 [cited 5 August 2014]. Available from: http://www.efpia.eu/uploads/EFPIA_Position_Paper_Off_Label_Use_May_2014.pdf. Accessed 8 August 2014.

224 Palmer E. Italy's move to fund unapproved use of Roche's Avastin alarms the industry [Internet]. FiercePharma. 2014 [cited 5 August 2014]. Available from: http://www.fiercepharma.com/story/italys-move-fund-unapproved-use-roches-avastin-alarms-industry/2014-06-11. Accessed 8 August 2014.

225 Kesselheim AS, Avorn J. Pharmaceutical promotion to physicians and first amendment rights. N Engl J Med. 2008;358(16):1727–32.

226 Anon. Office of the Commissioner C for DE and R. Guidances – [Internet] 2014 [cited 26 April 2014]. Available from: http://www.fda.gov/regulatoryinformation/guidances/ucm126486.htm. Accessed 8 August 2014.

227 Press BSM, Associated. Oklahoma House redefines abortion inducing drugs [Internet]. Washington Times. 2014 [cited 12 March 2014]. Available from: http://www.washingtontimes.com/news/2014/mar/10/oklahoma-house-redefines-abortion-inducing-drugs/?page=all. Accessed 8 August 2014.

228 Rosenbaum MB, Chiale PA, Halpern MS, *et al.* Clinical efficacy of amiodarone as an antiarrhythmic agent. Am J Cardiol. 1976;38(7):934–44.

229 Larkin I, Ang D, Avorn J, Kesselheim AS. Restrictions on pharmaceutical detailing reduced off-label prescribing of antidepressants and antipsychotics in children. Health Aff. 2014;33(6):1014–23.

230 Mello MM, Studdert DM, Brennan TA. Shifting terrain in the regulation of off-label promotion of pharmaceuticals. N Engl J Med. 2009;360(15):1557–66.

231 Kesselheim AS, Darby D, Studdert DM, *et al*. False claims act prosecution did not deter off-label drug use in the case of neurontin. Health Aff (Millwood). 2011;30(12):2318–27.

232 Kesselheim AS, Mello MM, Studdert DM. Strategies and practices in off-label marketing of pharmaceuticals: A retrospective analysis of whistleblower complaints. PLoS Med. 2011;8(4):e1000431.

233 Barber J. US Supreme Court rejects Pfizer's appeal over Neurontin marketing verdict [Internet] 9 December 2013 [cited 3 January 2014]. Available from: http://www. firstwordpharma.com/node/1170271. Accessed 8 August 2014.

234 Barber J. Pfizer agrees to pay $325 million in Neurontin marketing settlement [Internet] 2014 [cited 5 August 2014]. Available from: http://www.firstwordpharma.com/ node/1214430. Accessed 8 August 2014.

235 Singer N. Maker of Botox settles inquiry on off-label marketing. The New York Times [Internet]. 1 September 2010 [cited 23 January 2014]. Available from: http://www. nytimes.com/2010/09/02/business/02allergan.html. Accessed 8 August 2014.

236 Davis C, Abraham J. Is there a cure for corporate crime in the drug industry? BMJ. 2013;346(feb06 1):f755–f755.

237 Rémuzat C, Toumi M, Falissard B. New drug regulations in France: What are the impacts on market access? Part 2 – Impacts on market access and impacts for the pharmaceutical industry. J Mark Access Health Policy [Internet]. 6 August 2013 [cited 22 January 2014];1(0). Available from: http://www.jmahp.net/index.php/jmahp/article/view/20892/ 29571#CIT0005_20892. Accessed 8 August 2014.

238 Osborn J. Can I tell you the truth? A comparative perspective on regulating off-label scientific and medical information. Yale J Health Policy Law Ethics [Internet]. 2013;10(2): 299–356. Available from: http://digitalcommons.law.yale.edu/yjhple/vol10/iss2/2

239 Pharmaceutical promotion and the UK Bribery Act chapter – Pharmaceutical advertising 2013 [Internet]. International Comparative Legal Guides [cited 9 April 2014]. Available from: http://www.iclg.co.uk/practice-areas/pharmaceutical-advertising/pharmaceutical-advertising-2013/pharmaceutical-promotion-and-the-uk-bribery-act. Accessed 8 August 2014.

240 Fairman KA, Curtiss FR. Regulatory actions on the off-label use of prescription drugs: Ongoing controversy and contradiction in 2009 and 2010. J Manag Care Pharm. 2010;16(8):629.

241 Edwards P, Arango M, Balica L, *et al*. Final results of MRC CRASH, a randomised placebo-controlled trial of intravenous corticosteroid in adults with head injury-outcomes at 6 months. Lancet. 2005;365(9475):1957–9.

242 Zimmerman JJ. A history of adjunctive glucocorticoid treatment for pediatric sepsis: Moving beyond steroid pulp fiction toward evidence-based medicine. Pediatr Crit Care Med J Soc Crit Care Med World Fed Pediatr Intensive Crit Care Soc. 2007;8(6):530–9.

243 Anon. Signals for two newly approved drugs and 2010 annual summary. QuarterWatch [Internet]. 6 November 2011 [cited 9 April 2014]. Available from: https://www.ismp.org/ quarterwatch/pdfs/2010Q4.pdf. Accessed 8 August 2014.

244 Staton T. FDA weighs the free-speech case for off-label marketing [Internet]. FiercePharmaMarketing. 2014 [cited 5 August 2014]. Available from: http://www. fiercepharmamarketing.com/story/fda-weighs-free-speech-case-label-marketing/2014-05-21. Accessed 8 August 2014.

245 Klein DB, Tabarrok A. Do off-label drug practices argue against FDA efficacy requirements? A critical analysis of physicians' argumentation for initial efficacy requirements. Am J Econ Sociol. 2008;67(5):743–75.

246 Eichler H-G, Pignatti F, Flamion B, *et al*. Balancing early market access to new drugs with the need for benefit/risk data: A mounting dilemma. Nat Rev Drug Discov. 2008;7(10):818–26.

247 Eichler H-G, Bloechl-Daum B, Brasseur D, *et al*. The risks of risk aversion in drug regulation. Nat Rev Drug Discov. 2013;12(12):907–16.

248 US *et al.* ex rel. Lauterbach v. Orphan Medical Inc., Jazz Pharmaceuticals Inc., and Dr. Peter Gleason. Civil Action 05-CV-0387-SJF-KAM [Internet]. Feb 17, 2006 [cited 12 February 2014]. Available from: http://www.drugepi.org/downloads/downloads/OrphanMedical_Complaint1.pdf. Accessed 8 August 2014.

249 Hebert PC, Stanbrook M. Indication creep: Physician beware. Can Med Assoc J. 2007; 177(7):697.

250 Jones CW, Handler L, Crowell KE, *et al.* Non-publication of large randomized clinical trials: Cross sectional analysis. BMJ. 2013;347(oct28 9):f6104–f6104.

251 Doshi P. From promises to policies: Is big pharma delivering on transparency? BMJ. 2014;348(feb26 2):g1615–g1615.

252 Maglione M, Maher AR, Hu J, *et al.* Off-label use of atypical antipsychotics: An update [Internet]. Rockville (MD): Agency for Healthcare Research and Quality (US); 2011 [cited 27 January 2014]. Available from: http://www.ncbi.nlm.nih.gov/books/NBK66081/. Accessed 8 August 2014.

253 Sikirica V, Pliszka SR, Betts KA, *et al.* Comparative treatment patterns, resource utilization, and costs in stimulant-treated children with ADHD who require subsequent pharmacotherapy with atypical antipsychotics versus non-antipsychotics. J Manag Care Pharm. 2012;18(9):676–89.

254 Jefferson T, Jones MA, Doshi P, *et al.* Neuraminidase inhibitors for preventing and treating influenza in healthy adults and children. Cochrane Database Syst Rev. 2014;4:CD008965. DOI:10.1002/14651858.CD008965.pub4.

255 Muthuri SG, Venkatesan S, Myles PR, *et al.* Effectiveness of neuraminidase inhibitors in reducing mortality in patients admitted to hospital with influenza A H1N1pdm09 virus infection: A meta-analysis of individual participant data. Lancet Respir Med [Internet]. 2014;2(5):395–404 [cited 10 April 2014]. Available from: http://www.thelancet.com/journals/lanres/article/PIIS2213-2600(14)70041-4/abstract. Accessed 8 August 2014.

256 Butler D. Tamiflu report comes under fire. Nature. 2014;508(7497):439–40.

257 Emmerich J, Dumarcet N, Lorence A. France's new framework for regulating off-label drug use. N Engl J Med. 2012;367(14):1279–81.

258 Le Jeunne C, Billon N, Dandon A, *et al.* Off-label prescriptions: How to identify them, frame them, announce them and monitor them in practice? Thérapie. 2013;68(4):233–9.

259 Liang BA, Mackey T. Reforming off-label promotion to enhance orphan disease treatment. Science. 2010;327(5963):273–4.

260 Dresser R, Frader J. Off-label prescribing: A call for heightened professional and government oversight. J Law Med Ethics J Am Soc Law Med Ethics. 2009;37(3):476–86, 396.

261 Edersheim JG, Stern TA. Liability associated with prescribing medications. Prim Care Companion J Clin Psychiatry. 2009;11(3):115–9.

262 Rosqfl PM, Coleman DL. The case for legal regulation of physicians' off-label prescribing. Notre Dame Law Rev. 2011;86(2):649–91.

263 Fred Gebhart CE. OIG calls for diagnosis information on prescriptions [Internet]. Drug Topics. 2012 [cited 20 February 2014]. Available from: http://drugtopics.modernmedicine.com/drug-topics/news/modernmedicine/modern-medicine-news/oig-calls-diagnosis-information-prescriptions. Accessed 8 August 2014.

264 Vox F. Why Good Doctors Prescribe Off-Label Treatments – Pacific Standard: The Science of Society [Internet] 30 September 2013 [cited 10 February 2014]. Available from: http://www.psmag.com/navigation/health-and-behavior/good-doctors-prescribe-label-treatments-67071/. Accessed 8 August 2014.

265 DeMonaco HJ, Ali A, von Hippel E. The major role of clinicians in the discovery of off-label drug therapies. Pharmacother J Hum Pharmacol Drug Ther. 2006;26(3):323–32.

266 Why sharing data is for greater good. BBC [Internet]. 7 February 2014 [cited 10 February 2014]. Available from: http://www.bbc.co.uk/news/health-25988534. Accessed 8 August 2014.

267 Jung K, LePendu P, Chen WS, *et al.* Automated detection of off-label drug use. PLoS ONE. 2014;9(2):e89324.

268 Morris J. The use of observational health-care data to identify and report on off-label use of biopharmaceutical products. Clin Pharmacol Ther. 2012;91(5):937–42.

269 Weynants L, Schoonderbeek C. Off-label use and promotion in the EU: Risks and potential liability. Scrip Regulatory Affairs. 26 October 2010.

270 Mahmood K, Naeem M, Rahimnajjad NA. Metformin: The hidden chronicles of a magic drug. Eur J Intern Med. 2013;24(1):20–6.

271 Lee M-S, Hsu C-C, Wahlqvist ML, et al. Type 2 diabetes increases and metformin reduces total, colorectal, liver and pancreatic cancer incidences in Taiwanese: A representative population prospective cohort study of 800,000 individuals. BMC Cancer. 2011;11(1):20.

272 Chaudhry IB, Hallak J, Husain N, et al. Minocycline benefits negative symptoms in early schizophrenia: A randomised double-blind placebo-controlled clinical trial in patients on standard treatment. J Psychopharmacol Oxf Engl. 2012;26(9):1185–93.

273 Huang C-C, Chan W-L, Chen Y-C, et al. Statin use in patients with asthma: A nationwide population-based study. Eur J Clin Invest. 2011;41(5):507–12.

274 Klein BK, Klein R, Lee KE, et al. Statin use and incident nuclear cataract. JAMA. 2006;295(23):2752–8.

275 Frost FJ, Petersen H, Tollestrup K, et al. Influenza and COPD mortality protection as pleiotropic, dose-dependent effects of statins. Chest. 2007;131(4):1006–12.

276 Otte C, Zhao S, Whooley MA. Statin use and risk of depression in patients with coronary heart disease: Longitudinal data from the heart and soul study. J Clin Psychiatry. 2012;73(05):610–5.

277 Panagiotoglou TD, Ganotakis ES, Kymionis GD, et al. Atorvastatin for diabetic macular edema in patients with diabetes mellitus and elevated serum cholesterol. Ophthalmic Surg Lasers Imaging Off J Int Soc Imaging Eye. 2010;41(3):316–22.

278 Etminan M, Samii A, Brophy JM. Statin use and risk of epilepsy. Neurology. 2010;75(17):1496–500.

279 Marcus MW, Müskens RPHM, Ramdas WD, et al. Cholesterol-lowering drugs and incident open-angle glaucoma: A population-based cohort study. PloS One. 2012;7(1):e29724.

280 Stein JD, Newman-Casey PA, Talwar N, et al. The relationship between statin use and open-angle glaucoma. Ophthalmology. 2012;119(10):2074–81.

281 Hippisley-Cox J, Coupland C. Unintended effects of statins in men and women in England and Wales: Population based cohort study using the Qresearch database. BMJ. 2010;340(may19 4):c2197–c2197.

282 Vinogradova Y, Coupland C, Hippisley-Cox J. Risk of pneumonia in patients taking statins: Population-based nested case-control study. Br J Gen Pract J R Coll Gen Pract. 2011;61(592):e742–8.

283 Banaszewska B, Pawelczyk L, Spaczynski RZ, et al. Effects of simvastatin and metformin on polycystic ovary syndrome after six months of treatment. J Clin Endocrinol Metab. 2011;96(11):3493–501.

284 Marcella SW, David A, Ohman-Strickland PA, et al. Statin use and fatal prostate cancer. Cancer. 2012;118(16):4046–52.

285 Gutt R, Tonlaar N, Kunnavakkam R, et al. Statin use and risk of prostate cancer recurrence in men treated with radiation therapy. J Clin Oncol Off J Am Soc Clin Oncol. 2010;28(16):2653–9.

286 Kollmeier MA, Katz MS, Mak K, et al. Improved biochemical outcomes with statin use in patients with high-risk localized prostate cancer treated with radiotherapy. Int J Radiat Oncol Biol Phys. 2011;79(3):713–8.

287 Naseri M, Hadipour A, Sepaskhah M, et al. The remarkable beneficial effect of adding oral simvastatin to topical betamethasone for treatment of psoriasis: A double-blind, randomized, placebo-controlled study. Niger J Med J Natl Assoc Resid Dr Niger. 2010;19(1):58–61.

288 Abou-Raya A, Abou-Raya S, Helmii M. Statins: Potentially useful in therapy of systemic sclerosis-related Raynaud's phenomenon and digital ulcers. J Rheumatol. 2008;35(9):1801–8.

289 Chodick G, Amital H, Shalem Y, et al. Persistence with statins and onset of rheumatoid arthritis: A population-based cohort study. PLoS Med. 2010;7(9):e1000336.

290 Jick SS, Choi H, Li L, et al. Hyperlipidaemia, statin use and the risk of developing rheumatoid arthritis. Ann Rheum Dis. 2009;68(4):546–51.

291 Tang T-T, Song Y, Ding Y-J, *et al.* Atorvastatin upregulates regulatory T cells and reduces clinical disease activity in patients with rheumatoid arthritis. J Lipid Res. 2011;52(5): 1023–32.

292 Fogerty MD, Efron D, Morandi A, *et al.* Effect of preinjury statin use on mortality and septic shock in elderly burn patients. J Trauma. 2010;69(1):99–103.

293 Tleyjeh IM, Kashour T, Hakim FA, *et al.* Statins for the prevention and treatment of infections: A systematic review and meta-analysis. Arch Intern Med. 2009;169(18):1658–67.

294 Gupta R, Plantinga LC, Fink NE, *et al.* Statin use and sepsis events [corrected] in patients with chronic kidney disease. JAMA. 2007;297(13):1455–64.

295 Dobesh PP, Klepser DG, McGuire TR, *et al.* Reduction in mortality associated with statin therapy in patients with severe sepsis. Pharmacotherapy. 2009;29(6):621–30.

296 Craig TR, Duffy MJ, Shyamsundar M, *et al.* A randomized clinical trial of hydroxymethyl-glutaryl- coenzyme a reductase inhibition for acute lung injury (The HARP Study). Am J Respir Crit Care Med. 2011;183(5):620–6.

297 Li Y, Gottlieb J, Ma D, *et al.* Graft-protective effects of the HMG-CoA reductase inhibitor pravastatin after lung transplantation – A propensity score analysis with 23 years of follow-up. Transplantation. 2011;92(4):486–92.

298 Noël M, Gagné C, Bergeron J, *et al.* Positive pleiotropic effects of HMG-CoA reductase inhibitor on vitiligo. Lipids Health Dis. 2004;3:7.

299 Waltz E. It's official: Biologics are pharma's darlings. Nat Biotechnol. 2014;32(2):117.

300 Numedicus. Data exclusivity for biologics: No longer a level playing-field [Internet]. numedicus.co.uk [cited 27 April 2014]. Available from: http://numedicus.co.uk/blog/?p=195. Accessed 8 August 2014.

Index

Note: Page numbers in *italics* refer to Figures; those in **bold** to Tables.

Off-label Prescribing – Justifying Unapproved Medicine, First Edition. David Cavalla.
© 2015 John Wiley & Sons, Ltd. Published 2015 by John Wiley & Sons, Ltd.